INTERMEDIATE-LEVEL IMAGE PROCESSING

INTERMEDIATE-LEVEL IMAGE PROCESSING

Edited by

M. J. B. DUFF

Department of Physics and Astronomy, University College London, UK

1986

ACADEMIC PRESS
Harcourt Brace Jovanovich, Publishers
London   Orlando   New York   San Diego   Austin
Montreal   Sydney   Tokyo   Toronto

ACADEMIC PRESS INC. (LONDON) LTD
24/28 Oval Road, London NW1 7DX

United States Edition published by
ACADEMIC PRESS INC.
Orlando, Florida 32887

Copyright © 1986 by
ACADEMIC PRESS INC. (LONDON) LTD
All rights reserved. No part of this book may be reproduced
in any form by photostat, microfilm, or any other means,
without written permission from the publishers

British Library Cataloguing in Publication Data

Intermediate-level image processing
1. Image processing
I. Duff, Michael J.B.
006.4′2   TA1632
ISBN 0-12-223325-5

PRINTED IN GREAT BRITAIN AT THE ALDEN PRESS
OXFORD LONDON AND NORTHAMPTON

Contributors

H.G. BARTELS, *Optical Sciences Center, University of Arizona, Tucson, Arizona 85721, USA*
P.H. BARTELS, *Optical Sciences Center, University of Arizona, Tucson, Arizona 85721, USA*
J.-L. BASILLE, *Département Informatique, Institut Universitaire de Technologie, 50A Chemin des Maraîchers, 31062 Toulouse, France*
S.A. BOBMAN, *Department of Radiology, Box 3302, Duke University Medical Center, Durham, North Carolina 27710, USA*
V. CANTONI, *Dipartimento di Informatica e Sistemistica, University of Pavia, Via Abbriatgrasso 209, 27100 Pavia, Italy*
S. CASTAN, *Département Informatique, Institut Universitaire de Technologie, 50A Chemin des Maraîchers, 31062 Toulouse, France*
P. DALLE, *Département Informatique, Institut Universitaire de Technologie, 50A Chemin des Maraîchers, 31062 Toulouse, France*
R.N. DIXON, *Department of Medical Biophysics, University of Manchester, Stopford Building, Oxford Road, Manchester M13 9PT, England*
M.J.B. DUFF, *Department of Physics and Astronomy, University College London, Gower Street, London WC1E 6BT, England*
A. FAVRE, *Institute of Anatomy, University of Berne, Buehlstrasse 26, CH-3000 Berne 9, Switzerland*
P. GEMMAR, *Research Institute for Information Processing and Pattern Recognition FIM (FGAN e.V.), Eisenstockstrasse 12, 7505 Ettlingen 6, West Germany*
B.K. GILBERT, *Department of Physiology and Biophysics, Mayo Foundation, Rochester, Minnesota 55905, USA*
J. GRAHAM, *Department of Medical Biophysics, University of Manchester, Stopford Building, Oxford Road, Manchester M13 9PT, England*
P.J. GREGORY, *Visual Machines Limited, Enterprise House, Manchester Science Park, Lloyd Street North, Manchester M13, England*
W.G. GRISWOLD, *Optical Sciences Center, University of Arizona, Tucson, Arizona 85721, USA*
D. HILLMAN, *Optical Sciences Center, University of Arizona, Tucson, Arizona 85721, USA*

R. HUMMEL, *Department of Computer Sciences, Courant Institute of Mathematical Sciences, New York University, 251 Mercer Street, New York, New York 10012, USA*

L.H. JAMIESON, *School of Electrical Engineering, Purdue University, West Lafayette, Indiana 47907, USA*

C.H. JEON, *School of Electrical Engineering, Cornell University, Ithaca, New York 14853, USA*

J.T. KUEHN, *Institute for Defense Analyses, Supercomputing Research Center, 4380 Forbes Boulevard, Lanham, Maryland 20706, USA*

M. LAVIN, *IBM, T.J. Watson Research Center, Yorktown Heights, New York 10598, USA*

J.N. LEE, *Department of Radiology, Box 3302, Duke University Medical Center, Durham, North Carolina 27710, USA*

S. LEVIALDI, *Dipartimento di Matematica, University of Rome, Piazzale Aldo Moro 2, 00185 Rome, Italy*

H. LI, *IBM, T.J. Watson Research Center, Yorktown Heights, New York 10598, USA*

R. MAENNER, *Optical Sciences Center, University of Arizona, Tucson, Arizona 85721, USA*

M. NAGAO, *Department of Electrical Engineering, Kyoto University, Yoshidahon-machi, Sakyo-ku, Kyoto 606, Japan*

A.P. REEVES, *School of Electrical Engineering, Cornell University, Ithaca, New York 14853, USA*

S.J. RIEDERER, *Department of Radiology, Box 3302, Duke University Medical Center, Durham, North Carolina 27710, USA*

D.H. SCHAEFER, *Department of Electrical and Computer Engineering, George Mason University, Fairfax, Virginia 22030, USA*

R.L. SHOEMAKER, *Optical Sciences Center, University of Arizona, Tucson, Arizona 85721, USA*

H.J. SIEGEL, *PASM Parallel Processing Laboratory, School of Electrical Engineering, Purdue University, West Lafayette, Indiana 47907, USA*

Q.F. STOUT, *Department of Electrical Engineering and Computer Science, University of Michigan, Ann Arbor, Michigan 48109, USA*

S.L. TANIMOTO, *Department of Computer Science, FR-35, University of Washington, Seattle, Washington 98195, USA*

C.J. TAYLOR, *Department of Medical Biophysics, University of Manchester, Stopford Building, Oxford Road, Manchester M13 9PT, England*

L. UHR, *Computer Sciences Department, University of Wisconsin, Madison, Wisconsin 53706, USA*

C.-C. WANG, *IBM, T.J. Watson Research Center, Yorktown Heights, New York 10598, USA*

Preface

Operations on images can be grouped very roughly into two categories. In low-level processing the input data are the pixels of the original image and the output data represent properties of the image in the form of numerical values associated with each pixel. High-level processing is usually regarded as those parts of the processing in which the input is a set of derived features from the image and the output is an interpretation of the image content.

Expressed in this way, it is obvious that there may be a missing category in which the input is a set of values still associated with each pixel (i.e. when an n by n pixel image has produced an n by n array of values, or pixel labels) and the output is a list of features. This we are calling intermediate-level processing.

As it happens, not only does intermediate-level processing tend to be forgotten, but so also processor systems optimized for this level are largely non-existent. In practice, the intermediate-level operations are usually absorbed into either the level above or below.

At the seventh annual workshop in a series devoted to exploring the relationships between languages, architectures and algorithms for image processing, a group of some forty researchers met in the attractive and historic Chateau de Bonas, Gers, France, in May 1985. The special topic for discussion was Intermediate-Level Processing, and, as has been the custom at these workshops, an attempt was made to increase understanding not by formal presentation of prepared papers but rather by argument and exposure of sometimes half-formed ideas to an often aggressively critical audience. The nineteen chapters in this book were nearly all written after the workshop and can therefore be regarded as embodying at least some of the points which emerged in the discussions. However, this is not a "proceedings" nor is it a structured text on intermediate-level processing. No attempt has been made to edit out overlaps; duplicated material is retained so that each chapter can stand alone. On the other hand, some attempt has been made to shape all the manuscripts into a uniform format and, with the authors' permission and help, to modify some of the text to make it more comprehensible to a non-specialist reader.

The chapters have been grouped so as to bring together work on related topics, and each section is briefly summarized in an editorial introduction. Readers seeking more detailed information on any particular project are invited to correspond with the appropriate author, whose address, at the time of publishing, is listed in the front of this book.

Readers of this book may be interested to know that the earlier workshops in this series have also stimulated publications. The first workshop was held in Windsor in 1979 and was followed in 1980 by a meeting in Ischia. Participants in these two workshops contributed the material for *Languages and Architectures for Image Processing*, edited by M.J.B. Duff and S. Levialdi. Subsequent workshops in Madison (1981), Abingdon (1982), Polignano (1983) and Tucson (1984) resulted in *Multicomputers and Image Processing*, edited by K. Preston Jr. and L. Uhr, *Computing Structures for Image Processing*, edited by M.J.B. Duff, *Integrated Technology for Parallel Image Processing*, edited by S. Levialdi, and *Evaluation of Multi-Computers for Image Processing*, edited by K. Preston Jr., L. Uhr, S. Levialdi and M.J.B. Duff (now in press).

My personal thanks are due to all the contributors to this book and to all who took part in the discussions at Bonas. It should not go unrecorded that, without exception, every manuscript reached me on or before the deadline (give or take a few hours to allow for transatlantic time differences).

University College London   *Michael J. B. Duff*
October 1985

Contents

Contributors
Preface

INTERMEDIATE-LEVEL PROCESSING

Chapter One
Architectural issues for intermediate-level vision
    S.L. TANIMOTO   3

Chapter Two
An architecture for integrating symbolic and numerical image processing
    C.J. TAYLOR, R.N. DIXON, P.J. GREGORY and J. GRAHAM   19

Chapter Three
Shape recognition by human-like trial-and-error random processes
    M. NAGAO   35

PARALLEL ALGORITHMS

Chapter Four
The mapping of parallel algorithms to reconfigurable parallel architectures
    L.H. JAMIESON   53

Chapter Five
The V-language for polymorphic architectures and algorithms
    H. LI, C.-C. WANG and M. LAVIN   65

Chapter Six
Considerations on parallel solutions for conventional image algorithms
    P. GEMMAR   83

Chapter Seven
Connected component labelling in image processing with MIMD architectures
    R. HUMMEL   101

PYRAMIDS

Chapter Eight
Multiple-image and multimodal augmented pyramid computers
    L. UHR   131

Chapter Nine
Algorithm-guided design considerations for meshes and pyramids
    Q.F. STOUT   149

Chapter Ten
Pyramid architectures
    D.H. SCHAEFER   167

Chapter Eleven
Contour labelling by pyramidal processing
    V. CANTONI and S. LEVIALDI   181

SYSTEM STUDIES

Chapter Twelve
Computer-vision task distribution on a Multicluster MIMD system
    A.P. REEVES and C.H. JEON   193

Chapter Thirteen
Multifunction processing with PASM
    J.T. KUEHN and H.J. SIEGEL   209

Chapter Fourteen
Iconic and symbolic use of a line processor in multilevel structures
    J.-L. BASILLE, P. DALLE and S. CASTAN   231

Chapter Fifteen
Self-learning capabilities of VAP for low-level vision
    A. FAVRE   243

Contents

Chapter Sixteen
The impact of the emerging gallium arsenide integrated-circuit technology on algorithms and computer architectures for signal and image processing
    B.K. GILBERT   253

APPLICATIONS

Chapter Seventeen
Multiprocessor computer system for medical image processing
    W.G. GRISWOLD, P.H. BARTELS, R.L. SHOEMAKER, H.G. BARTELS, R. MAENNER and D. HILLMAN   267

Chapter Eighteen
Processing techniques for magnetic resonance image synthesis
    S.J. RIEDERER, J.N. LEE and S.A. BOBMAN   287

CONCLUSION

Chapter Nineteen
Complexity
    M.J.B. DUFF   307

Subject Index   315

INTERMEDIATE-LEVEL PROCESSING

In this first section, Steven Tanimoto sets the scene by discussing the general problem of the missing architectural layer in image processing, i.e. the intermediate level, and notes that whereas attention has been given to the design of special architectures for low-level processing (particularly by SIMD arrays) and high-level processing (usually by multiple serial processors), the middle layer has been largely neglected. Christopher Taylor and collaborators then describe a system that attempts to bridge this gap, starting at the higher level and then adding specialized processors to improve low-level performance, paying particular attention at the same time to the choice of appropriate data structures. The use of well-chosen data structures in the intermediate level is emphasized by Makoto Nagao, whose approach is based on the application of "slits", which focus attention on parts of the image requiring detailed analysis. Taylor also regards his data structures as a way of focusing attention on important image detail, and it seems that this aspect of intermediate-level processing may be an important factor in narrowing the low-level, high-level gulf.

Chapter One

Architectural Issues for Intermediate-Level Vision

S. L. TANIMOTO

1 INTRODUCTION

1.1 Low-level and high-level vision

There has been a great improvement in our understanding of the proper role of special hardware for image-processing operations at the pixel level during the last decade. The CLIP4 and MPP systems were successfully developed, offering a 96 by 96 array and 128 by 128 array of processing elements, respectively. The essential characteristics of these iconic processors are these: they provide one processing element for each pixel of an image; the processing elements are connected in a two-dimensional mesh; there is a single control unit for all of the processing elements, so that the system falls into the Single-Instruction-stream, Multiple-Data-stream classification (SIMD). Two additional aspects of the CLIP4 and MPP systems are that (1) their processing elements operate in a bit-serial way, and (2) at any instant the addresses used by the processing elements to access their local memories are all identical. The iconic processors are typically used to filter images, detect edges of objects, and sometimes to extract useful global features such as the mean pixel value. (For a description of typical CLIP4 applications see [1]; the Massively Parallel Processor is described in [2].) One could say that the "responsibility" of an iconic processor, as a component of a computer-vision system, is to transform sensed images into ones that are more useful for understanding the contents of the image (i.e. the scene).

There has also been a significant improvement in computer architecture for symbolic computation. For example the Symbolics 3600 series of LISP machines has been developed and marketed. These systems are appropriate for manipulating the semantic structures needed in scene understanding, where contextual knowledge plays a large role. The "responsibility" of a symbolic processor in a computer-vision system is to take two kinds of information: (1) a partial description of a scene, derived from the image, and (2) world knowledge in the form of constraints, schemata, and/or production rules; and to produce as complete a scene description as necessary for the application. In a computer system to perform general vision tasks such as those handled by humans (see [3]), there is a clear need for efficient architecture at both the image level and the symbolic level of processing.

1.2 The intermediate level

Largely missing from the field today, however, are architectures that are suitable for rapidly transforming image data into symbolic form. The image-array architectures, while highly efficient for processing images, are suddenly bottlenecked when they have to deal with lists and symbols. On the other hand, the LISP machines currently in use for work in artificial intelligence just do not have the massive parallelism needed for image processing in real time. This paper is concerned with the missing architectures. Why are they needed? What could they be like?

It is convenient to use the term "intermediate-level vision" to refer to the processing of information that takes place between the "low level" of image processing and the "high level" of semantic analysis. The meaning of the term is further refined as follows: a transformation of intermediate-level vision is one that takes as input an image represented as a two-dimensional array, and outputs a structure that is *not* a two-dimensional array. While this definition has the disadvantage that it permits some strange and arbitrary transformations to be classified as iconic-to-symbolic, it serves the purpose of distinguishing transformations such as chain encoding, shape measurement, and building a region-adjacency graph from image-processing operations such as median filtering and convolution. It also has the advantage of being a simple definition.

1.3 Retinotopic and nonretinotopic architectures

Referring to the human visual system, neurophysiologists make a distinction between "retinotopic" levels of the visual cortex and "nonretinotopic"

Architectural issues

levels. A retinotopic area of the brain is one in which the visual signals carried by the neurons have the spatial structure that they do in the retina. That is, the retinal activity is "conformally mapped" into the neural matter in a retinotopic level, even though the signals may represent enhancements of the retinal responses. By contrast, a nonretinotopic region is one in which there is no longer any one-to-one and continuous correspondence between small areas of the retina and small areas in the cortical level.

The fact that this difference in structure between the retinotopic and nonretinotopic areas of the brain exists lends support to the position that powerful machine-vision systems ought themselves to incorporate both iconic parallel hardware and semantic-network hardware.

The question immediately arises of how the iconic hardware is to be coupled to the semantic hardware. It has been estimated that in the human visual system, information flows across the retinotopic/nonretinotopic interface at the rate of 10 billion bits per second, over some one billion connections. Neither currently available image processors nor symbolic processors can input or output data at such rates. Thus, even if one does not want to provide special hardware for computing intermediate-level vision transformations, one needs a new kind of interface simply to allow the image-level hardware to communicate effectively with the semantics-level hardware.

There are therefore two approaches to the problem of hardware for the intermediate level of vision: (a) special devices to compute iconic-to-symbolic and/or symbolic-to-iconic transformations, and (b) high-speed interfaces between the low-level (image-oriented) computing systems and the high-level (symbol-oriented) systems. Naturally, these approaches do not exhaust the possibilities—one could design hardware that could perform both the transformation and interface functions, for example. However, the two approaches provide a framework in which to explore possible designs for hardware at the intermediate level of processing.

1.4 Cutting up the cake

In building a general machine-vision system such as VISIONS [3], one finds a need for computing hardware at all three of the levels previously mentioned. Past research and development efforts have made currently available the hardware for the low-level and the high-level parts of the problem, leaving us with the intermediate level uncovered (see Fig. 1).

For some applications of machine vision, the lack of intermediate-level hardware is not a problem, because complex symbolic descriptions of an image are not needed. To illustrate: a change-detection system that aids

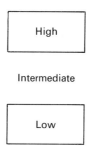

Fig. 1. *The current state of appropriate computer architecture. Intermediate-level vision is not covered.*

night-watchmen needs no sophisticated scene-description abilities. However, for complex visual tasks such as the visual control of a robot navigating over rugged terrain, there is clearly a problem.

There are several possible ways to solve this problem of covering the intermediate level. One of these is to enhance the capabilities of a (low-level) image processor so that it can handle iconic-to-symbolic transformations with ease. This is a "bottom-up" approach. It is illustrated in Fig. 2(a). The approach is exemplified by the idea of augmenting a pyramid machine [4,5], itself a hierarchical structure that may be considered to be an augmented cellular array [6,7]. An approach complementary to the bottom-up approach is to enhance the capabilities of a (high-level) symbol processor so that it is not swamped by the individual pixel accesses and computations needed for iconic-to-symbolic transformations. This, the "top-down" approach, is shown in Fig. 2(b).

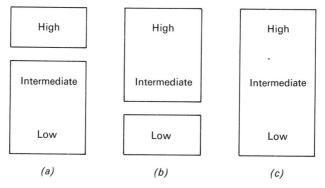

Fig. 2. *(a) Bottom-up approach to covering the intermediate level; (b) top-down approach; and (c) unified approach.*

Architectural issues

Some computer architects feel that there exist single architectures appropriate to all three levels of processing. The architectures they suggest are usually multiprocessor MIMD systems that are reasonably appropriate for high-level work but relatively inefficient for low-level work when compared to the CLIP and MPP systems. (For a survey of both mesh-connected SIMD and general MIMD network architectures see [4].) Reconfigurable architectures such as the PASM [8] have the possibility of covering the entire problem; however, the provision of a large enough number of processing elements to reach MPP comparability in SIMD mode would appear to be very expensive.

There is one approach that takes the view that some computational tasks of machine vision are inherently intermediate, and so, special hardware should be provided at the intermediate level. The inherently intermediate computation approach is shown in Fig. 3. With this idea, one provides some

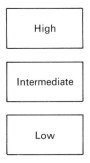

Fig. 3. The "inherently-intermediate" approach. Each level has its own specialized architecture.

kind of hardware just for the intermediate level. The hardware might be general—capable of many different iconic-to-symbolic and symbolic-to-iconic transformations—or could be very specific, computing only one such transformation (say, a form of the Hough transform).

Even if one takes the bottom-up or top-down approach, one is faced with the problem of what the enhancements to the existing low-level or high-level processor should be, and this problem is not necessarily greatly different from designing a completely new computing module.

In the rest of this paper, we take the view that some new hardware is needed to cover the intermediate level and that there exist inherently intermediate computations in machine vision.

2 WHAT ICONIC–SYMBOLIC ALGORITHMS ARE SUITABLE FOR SPECIAL HARDWARE?

We have distinguished two approaches to filling the void at the intermediate level: one is to provide hardware specifically to compute iconic-to-symbolic and possibly symbolic-to-iconic transformations. The other is to provide an interface with adequate bandwidth to allow the iconic and symbolic processes to communicate without a bottleneck. In this section we consider the first approach, hardware for iconic/symbolic transformations.

2.1 Three classes: descriptions of points, contours or regions

What are the most important iconic-to-symbolic transformations? Although this must depend on the applications and algorithms one wishes to use, some general evaluations can be made. Most symbolic descriptions of images can be built from descriptions of points, contours (including lines and edge sequences) and regions. Transformations that create representations of these objects or of their properties are good candidates for implementation in special hardware. (They are also good candidates for implementation in algorithms that work through special high-level/low-level interfaces.)

Certain kinds of points are particularly valuable in image description. These include endpoints of lines and curves, and points of high curvature (e.g. corners) or inflection; points of maximum or minimum brightness over an area; centres or centroids of regions. Associated with such points may be properties such as the directions of incident lines, the curvature of the incident curve, the brightness of the pixel at the point, the texture in the vicinity of the point, etc. Algorithms that identify and extract the coordinates and/or properties of such points from images are good examples of iconic-to-symbolic transformations.

Prominent lines or contours in an image may represent object boundaries, texture edges or other events. An operation that identifies from an image one or more lines or curves and creates a (non-image) representation of the lines or curves and/or their properties is an iconic-to-symbolic transformation. Thus, for example, the detection of closed edge contours and the measurement and output of their parameters is such a transformation.

Regions are usually defined to be connected groups of pixels. However, they could be defined more generally as subsets of the image plane having nonzero area. Regions are the typical results of segmentation processes. If an operation produces from an image a non-image representation of regions and/or properties of regions, it is also an iconic-to-symbolic transformation. An operator that produces an average texture value (e.g. coarseness) for an

Architectural issues 9

entire image is an iconic-to-symbolic transformation of this type, since the value is a property of the whole image, which is a region.

Algorithms that compute iconic-to-symbolic transformations such as these are usually computationally intensive. By having to examine large numbers of pixels the algorithms are destined to take a significant time. If most of the pixel-accessing activity can be handled by the iconic (low-level) processor, then the task of the special intermediate-level hardware can be simplified. For example, an algorithm that produces descriptions of corner points from a binary image could employ an iconic processor to detect the corners using a template-matching approach, and then the special hardware could construct explicit representations of the coordinates of the corners; the special hardware could consist of several processors, each of which scans a portion of the processed image in which a 1 indicates the presence of a corner in a particular orientation.

The output of an iconic-to-symbolic transformation is typically not a single object (e.g. a single point or a single region) but a set or list of many such objects. These objects generally are distributed over the image, but seldom in a uniform distribution. As such, if processors or processing elements for doing the intermediate-level transformations are hardwired to handle fixed areas of the image, then the load on them will tend to be nonuniform. Furthermore, except in the case of describing points, the descriptions of objects (lines, curves, regions) will themselves vary in size and vary in the time required to compute them, further complicating load balancing and synchronization. Even more disruptive is the fact that individual lines, curves or regions may extend across an image so as to fall into the zones of several processors; this can require that processors coordinate their activities to form descriptions of complete lines, curves or regions.

2.2 Hough transforms

Because the Hough transform is concerned with finding and describing lines in images, it is a good candidate for implementation in special, intermediate-level, hardware. In its simplest form, the Hough transform is computationally expensive, even for most parallel computers. However, the method is easily modified in several ways that make it amenable to more efficient computation.

The standard Hough transform of an image $F(x, y)$ is an array of values $A(\rho, \theta)$, where

$$A(\rho,\theta) = \sum_{x=1}^{N} \sum_{y=1}^{N} F(x,y) h(x,y,\rho,\theta),$$

where $h(x,y,\rho,\theta)$ is a "hit function" or incidence function defined as

$$h(x,y,\rho,\theta) = \begin{cases} 1 & \text{if } |x\cos\theta + y\sin\theta - \rho| < \varepsilon, \\ 0 & \text{otherwise.} \end{cases}$$

That is, (x, y) is judged to hit the line $\rho = x\cos\theta + y\sin\theta$ if its distance from the line is within ε.

The array $A(\rho, \theta)$ is sometimes called an array of accumulators because each double sum is usually computed by an iterative process of adding new values to each element of the array. If F is a binary image, then each nonzero pixel "votes" for the pairs of parameters ρ and θ that specify lines that it hits. When F is a grey-value image the votes are weighted by each pixel value. If $A(\rho_0, \theta_0)$ has a relatively large value then this is an indication that the line $\rho_0 = x\cos\theta_0 + y\sin\theta_0$ has received strong voting and is one of the prominent lines in the image.

The following are some ways to modify the Hough transform that reduce its computational burden:

(i) Reduce the number of distinct values of ρ and θ considered in the parameter space. The result of this is that the number of different lines that can possibly be found is reduced. Each of these lines must then represent a range of lines; that is, each pair (ρ, θ) for which a vote total is accumulated must represent a rectangular region in the (ρ, θ) parameter space.

(ii) Reduce the resolution of the image. One way of accomplishing this is to not permit individual pixels to vote but to permit only blocks of pixels to vote.

(iii) Apply the Hough transform locally instead of globally in the image. For example, instead of computing one big Hough transform, compute many small ones, e.g. over 8 by 8 neighbourhoods. These small Hough transforms should use technique (i), the reduced parameter space.

(iv) Assume that an orientation is computed at each pixel of the image; this is the orientation of the dominant edge or line passing through the pixel. Restrict the Hough transform to only accumulate votes from pixels with a particular orientation. The parameter space may then be collapsed to a one-dimensional space and much less voting and vote-counting need be done.

Once a reduced form of the Hough transform has been agreed upon, one may proceed to design special hardware for its computation. The bottleneck is generally accumulating all the votes. One suggestion is that a special VLSI chip be fabricated to handle most of the vote accumulation [9]. If the counting is to be truly parallel, there is a problem of handling large fan-out

Architectural issues 11

and fan-in on the lines that distribute the votes. A combination of parallel and sequential vote counting may prove to be most economical in the final analysis.

The many ways in which the Hough transform may be adapted suggest that many different hardware and software solutions to the line-finding problem are possible, and it is very difficult to say in advance which would be best.

2.3 Connectivity analysis

Another problem which fits neatly into the iconic-to-symbolic category is that of building a representation of a region that takes into account the entire set of pixels connected together in the region and none of those which are not connected to the region. While the input in this situation is an image, the output may be a list of the properties of each connected component in the image: the number of pixels constituting the region, the vertices of the bounding box around the region, etc.

Unlike the Hough transform, a connectivity analysis seems to require a certain amount of sequential processing in an image. To prove that a point P is connected to a point Q, one must show that there is a path from P to Q that stays within the designated region. In order to affix a label to P indicating that it belongs to the same region that Q does, that label must somehow propagate from Q to P, or there must be some other source Q', from which both P and Q receive the label. The labels must travel along the paths or along equivalent paths through data structures outside the image.

However, an algorithm for connectivity analysis can be designed to take advantage of the structure usually present in images [10]. Simple scanning methods work well when the region is convex, and the performance of scanning methods degrades gracefully as the shape of the region increases in complexity.

It is currently unknown whether reasonable hardware could be designed that would bring parallelism to the task of identifying and describing the connected components of a multivalued image. The importance of connectivity analysis suggests that more work should be done on this problem.

3 PROPOSED DEVICES FOR ICONIC/SYMBOLIC TRANSFORMATIONS

Two devices that have been proposed for computing iconic-to-symbolic transformations are the "ISMAP" (Iconic-to-Symbolic MAPper) [11], and a "chain-run" encoding device to accelerate the collection of chain curves from an image [12]. Each of these devices is inherently serial and requires that pixel data be sequentially fed to it.

3.1 Iconic-to-Symbolic MAPper

The ISMAP, to be used in conjunction with the PIPE (Pipelined Image Processing Engine [13]), is designed to rapidly compute a histogram or to use the histogram to build a table in which the coordinates of pixels of the original image are indexed according to pixel value. Thus, for example, one may easily construct a list of the coordinates of all the pixels that have value 27. With ISMAP, such an indexed structure could be produced in three video frame times (about 100 ms). This capability allows an algorithm designer considerable flexibility in devising procedures for tasks such as line finding. An algorithm that employs the PIPE, ISMAP and a host machine to efficiently find the dominant lines in an image has been devised and provides an attractive alternative to computing a full Hough transform and extracting peaks from it [14]. An option for ISMAP is under consideration that would allow it to run backwards, taking lists of coordinates and plotting particular values into the pixels specified. Such an option could be employed to produce "hypothesis images" used to bias or restrict image-processing algorithms in certain areas of an image.

3.2 Chain-run encoder

The chain-run encoding device [12] was designed to attach to an edge of a "systolic cellular logic" processor [15] and to input an entire row or column of a binary image at one time. The device rapidly scans the buffered row or column and outputs a list of the coordinates at which 1s are found. In conjunction with the cellular logic processor, it can provide the endpoints of runs in the chain-code representation of a binary image. These endpoints can be organized into vectors and sequences of vectors by a collection of microprocessors. The design suggests that one encoding device be used for each direction in the chain-coding system (i.e. four or eight). A custom VLSI chip contains logic that makes scanning the buffered row or column extremely rapid if the data is sparse, which it normally would be.

Architectural issues 13

4 INTERFACES

Rather than provide hardware for specific iconic-to-symbolic transformations, it may be sufficient to provide an interface that will allow many general-purpose processes (which would otherwise perform high-level tasks) to get at the pixels of the low-level section with high bandwidth. Then, provided there are enough processors, the iconic-to-symbolic transformations may be computed quickly enough.

4.1 Local or global access to the iconic array

Suppose that there are M symbolic processors to be interfaced to an iconic processor that includes an N by N array of processing elements. If M is a power of 4 and N is a power of 2, and $M < N^2$, then each symbolic processor may be assigned to a distinct block of iconic processing elements. Such an assignment implies that only the members of the block are capable of communicating with their assigned symbolic processor at the maximum speed. Other processing elements would have to route their information into the assigned block to reach the symbolic processor, or alternatively route it through other symbolic processors. If, for reasons of speed, a symbolic processor is effectively restricted to a small block of the iconic array, it is only capable of computing descriptions of objects that lie entirely within the block, without incurring a speed penalty. This approach of scattering the symbolic units across the image effectively imposes retinotopic structure onto the symbolic level of hardware.

On the other hand, if each of the M symbolic processors is given direct access to all N^2 iconic elements, there are at least two problems: (1) the cost of the system would be higher because of the increased number of connections needed, and (2) there could be contention among the M symbolic processors, for access to the iconic memory. With no spatial restrictions, the M symbolic processors would form a nonretinotopic arrangement. The principal tradeoff here is increased cost for greater globality of access to image data.

An example of a network of symbolic processors intended for use in a retinotopic arrangement is the "Semantic Network Array Processor" (SNAP) [16].

4.2 Memory-based interfaces

Let us assume that a nonretinotopic arrangement of symbolic processors is desired. That is, each symbolic processor is to have equally rapid access to all the pixels of the iconic array. One of the problems mentioned above is the possibility of contention by several symbolic processors for the same pixel. This problem can be alleviated by replicating the image data and letting each symbolic processor have its own copy or version. In order to store these extra copies, additional memory is required. An interface based around memory for storing such images can be called "memory-based".

One proposal for an interface between the iconic and symbolic levels is to use a large memory buffer which supports two kinds of access: by images and by pixel coordinates [17]. Such a memory may be called a "bimodal memory". The low-level system may read or write a binary image in the bimodal memory in one step. Meanwhile, several general-purpose microprocessors may be reading or writing 8-bit pixels in several different layers of the bimodal memory.

The efficient construction of the bimodal memory requires an unusual component: a memory chip that provides access both in terms of binary images and in terms of pixel coordinates. The VRAMs that have recently come onto the market support raster-graphic displays well, but their image-access mode is serial rather than parallel; this would greatly slow down the interface, and what may be worse, require the parallel image processor to spend a great deal of its time in input and output operations.

5 CONTROL OF AND BY THE INTERMEDIATE LEVEL

In introducing hardware for the intermediate level of visual-information processing, the overall control of the system may become more complicated. If the three subsystems are to work asynchronously, software protocols between them take most of the responsibility for coordinating their activities. How autonomous should the intermediate level be? Should it take orders from one or both of the other levels? Should it direct the activities of the other levels?

If, even with new hardware, the intermediate level is still the bottleneck for the whole system, then perhaps the intermediate level should be given primary control of the low- and high-level components. In that way, it can coordinate the production of image data and the consumption of symbolic descriptions so as to maintain an efficient flow through the intermediate level. This strategy of putting control at the weakest link in the process is analogous to letting the critical path in a PERT chart indicate the allocation of project resources.

While general control of the entire system's activities (e.g. controlling which algorithms are being executed at each level) may either be centralized

Architectural issues

or distributed, the control of access to particular data items (i.e. the resolution of concurrent access requests) is necessarily distributed for the sake of efficiency. The form of these relatively mundane control systems can be considered to be an implementation issue; interrupts, flags and message passing are some of the mechanisms at the designer's disposal.

If the intermediate-level hardware can easily keep up with the iconic and symbolic processors then overall control should probably not be at the intermediate level, central as it is. Presumably, most of the knowledge about what is being sought in the image resides in the symbolic level of the system, and this knowledge should be used in the control of the entire system, guiding its perception effort. There is thus a "knowledge argument" in favour of putting overall control at the symbolic level of the system. With control at the top, a system is well prepared to employ top-down strategies for machine vision.

The bottom-up strategy may be justified when there are structures in the image that are so compelling as to force particular interpretations, no matter what scenes or patterns were expected. Overall control might reasonably be put at the iconic level when pixel-oriented operations are the chief culprits, holding back the overall performance of the system and/or when not much flexibility is needed at the upper levels.

The idea of decentralizing the overall control has a certain appeal. There seem to be situations in which image contents should drive the system and other situations in which world knowledge should drive it. Decentralized control might make the system flexible enough to use the appropriate strategy in each situation. Such control mechanisms, however, are harder to design and understand, and probably harder to debug and maintain than centralized ones. If a hardware system is provided that would support alternative control strategies, experimental research could be performed that would lead to a better understanding of this issue.

6 CONCLUDING REMARKS

6.1 Brief summary of the issues raised

In order to build powerful, flexible machine-vision systems at a reasonable cost, new hardware to handle the intermediate level of visual-information processing must be designed and built, and new algorithms must be designed to make use of it. Exactly what class of operations the intermediate-level hardware should support is one issue. Another issue is whether the hardware should provide computing capability directly, or provide an efficient interface between the iconic and symbolic levels that will allow their computing capabilities to be fully utilized. Yet another issue is the extent to which control of the system should be given to the intermediate-level hardware. Designers of general-purpose machine-vision systems must resolve these issues to produce cost-effective systems.

6.2 Prospects for the future

It is likely that almost all of the alternatives mentioned here, for each design question, will be tried by some research and development group in the next five years; the needs of applications and the viewpoints of designers are diverse. There is commercial appeal to the idea of having plug-in modules for particular iconic-to-symbolic transformations, for example chain-encoding, connected component description, Hough transforms, region-adjacency graphs. At the same time, the idea of coupling the iconic and symbolic processing levels through high-bandwidth interfaces offers the possibility of high performance and generality—such an interface would permit almost any iconic-to-symbolic transformation to be computed more rapidly than is possible with a conventional (e.g. 32-bit bus) interface.

Acknowledgment

This research was supported in part by NSF Grant DCR-8310410.

REFERENCES

[1] Duff, M. J. B. (1985). Real applications on CLIP4. In [18], pp. 153–165.
[2] Batcher, K. E. (1980). Design of a massively parallel processor. *IEEE Trans. Comp.* **29**, 836–840.
[3] Hanson, A. R. and Riseman, E.M. (1978). VISIONS: A computer system for interpreting scenes. In *Computer Vision Systems* (ed. A. R. Hanson and E. M. Riseman), pp. 303–333. Academic Press, London and Orlando.
[4] Uhr, L. (1984). *Algorithm-Structured Computer Arrays and Networks: Architectures and Processes for Images, Percepts, Models, Information.* Academic Press, London and Orlando.
[5] Stout, Q. F. (1985). Mesh and pyramid computers inspired by geometric algorithms. In *Proc. Workshop on Algorithm-Guided Parallel Architectures for Automatic Target Recognition, Leesburg, Virginia,* 16–18 July 1984 (ed. C. L. Giles and A. Rosenfeld), pp. 293–315.
[6] Tanimoto, S. L. (1983). A pyramidal approach to parallel processing. In *Proc. 10th Int. Symp. on Computer Architecture, Stockholm, June 1983,* pp. 372–378.

[7] Tanimoto, S. L. (1984). A hierarchical cellular logic for pyramid computers. *J. Parallel and Distributed Comp.* **1,** 105–132.
[8] Kuehn, J. T., Siegel, H. J., Tuomenoksa, D. L. and Adams, G. B. (1985). The use and design of PASM. In [18], pp. 133–152.
[9] Ballard, D. H. and Brown, C. M. (1985). Visions. *Byte* **10,** 245–261.
[10] Danielsson, P.-E. and Tanimoto, S. L. (1983). Time complexity for serial and parallel propagation in images. In *Architecture and Algorithms for Digital Image Processing* (ed. A. Oosterlinck and P.-E. Danielsson). Proc. SPIE 435, August, pp. 60–67.
[11] Kent, E. W. (1985). Personal communication.
[12] Pfeiffer, J. (1985). Unpublished presentation at an informal workshop on pyramidal image processing systems, New Mexico State University, Las Cruces, NM, 12–13 April.
[13] Kent, E. W., Shneier, M. O. and Lumia, R. L. (1985). PIPE (pipelined image processing engine). *J. Parallel and Distributed Comp.* **2,** 50–78.
[14] Tanimoto, S. L. and Kent, E. W. (1986). *Architecture and Algorithms for Iconic/Symbolic Transformations.* University of Washington, Department of Computer Science, Technical Report No. 86-07-07, July 1986.
[15] Tanimoto, S. L. and Pfeiffer, J. J. (1981). An image processor based on an array of pipelines. In *Proc. IEEE Comp. Soc. Workshop on Computer Architecture for Pattern Analysis and Image Database Management, Hot Springs, Virginia, 11–13 November,* pp. 201–208.
[16] Moldovan, D. I. and Tung, Y. W. (1985). SNAP: A VLSI architecture for artificial intelligence processing. *J. Parallel and Distributed Comp.* **2,** 109–131.
[17] Tanimoto, S. L. (1985). An approach to the iconic/symbolic interface. In [18], pp. 31–38.
[18] Levialdi, S. (ed). (1985). *Integrated Technology for Parallel Image Processing.* Academic Press, London and Orlando.

Chapter Two
An Architecture for Integrating Symbolic and Numerical Image Processing

C. J. Taylor, R. N. Dixon, P. J. Gregory and J. Graham

1 INTRODUCTION

There are commonly held to be two, qualitatively distinct, types of activity involved in image processing. Low-level tasks require relatively simple processing, are predominantly numerical and involve large quantities of data. High-level processing is predominantly symbolic and involves relatively small quantities of data. It is often suggested that some form of special hardware is required for low-level processing and that this can be interfaced to a fairly conventional processor capable of undertaking high-level processing [1–3]. In this paper we suggest that it is important to take a more integrated approach to the problem. We confine our attention to relatively conventional computational methods and describe how our ideas have led to the design of two similar hardware systems, both of which have been commercially exploited. It must be said that there are more radical approaches to the problem of dealing with both numerical and symbolic processing, such as those proposed by advocates of connectionist machines [4,5]. These ideas are, however, a long way from practical implementation, and we are particularly interested in systems capable of undertaking visual tasks of useful complexity within the constraints of current technology.

2 COMBINING HIGH-LEVEL AND LOW-LEVEL PROCESSING

When an image is first read into an image-processing system it has very limited symbolic content and consists mainly of unstructured numerical data. In virtually all practical applications of image processing the aim is to generate a symbolic representation (interpretation) of the image. This may often be associated with a relatively small quantity of numerical data. There are normally additional data of both kinds available to the system (prior knowledge) which can be used in generating the desired solution. The processing that is close to the original image and which predominantly involves numerical data is often called low-level. That which is close to the solution and predominantly involves symbolic data is often called high-level. This distinction is useful for the purposes of discussion, although in practice the aim of any well-designed system is to present the programmer with an integrated set of representational methods that avoid any such dichotomy. The question that we address here concerns the relationship that should exist between the two types of processing.

2.1 The relationship between high-level and low-level processing

First we should make a point that is rather obvious but sometimes seems to be ignored. This is simply that when we attempt to automate real visual tasks it is generally the case that both kinds of processing play an important role. It is often recognized that the quantity of numerical data representing an image results in a large computational burden for low-level processes, which are, of necessity, applied to the whole image. It is not often recognized in this context that high-level processing can be considered to require an infinite computational effort. This can be argued by pointing out that there are, for visual tasks of realistic complexity, a virtually infinite number of possible symbolic interpretations of any image. The purpose of high-level processing is to search through these interpretations for one that is in some sense best. In a practical system it is obviously necessary to limit the search space by using heuristics. In the extreme this can lead to an approach where the symbolic element of the processing task is trivial compared with the numerical part. This is, however, an unsound approach, since heuristics that reliably make such dramatic reductions in the solution search space while still guaranteeing to retain the "correct" solution are not known. For these reasons we argue that it is important to apply as much computational effort as possible to high-level processing.

A possible response to this view is to accept the importance of high-level processing and argue that the requirement is thus for two different types of processor, one optimized for low-level processing and one, of conventional design, to undertake high-level processing. This ignores the fact that there is

Integration of symbolic and numerical processing

a rather intimate relationship between symbolic and numerical data which needs to be reflected in considerable commonality between the hardware involved in each type of processing. If we take a bottom-up approach to scene interpretation such as that proposed by Marr and followers [6,7] we will generate a hierarchy of structures, each of which have both symbolic and numerical content. The original image is almost purely numerical, but intrinsic images generated by low-level processing also have significant symbolic content. Primitives extracted from intrinsic images have greater symbolic content but still have significant numerical content and the same is true of the $2\frac{1}{2}$D sketch. The current consensus seems to accept that an arrangement such as this where information flows only in one direction may not be desirable and that the results of higher-level processes probably need to feed back to lower-level processes. This further reinforces the argument that high-level and low-level processing are intimately related and that this should be reflected in hardware architecture.

We have concluded from these arguments that it is the interface between high-level and low-level processing that is crucial. Whatever computing power may be provided independently for the two types of activity, system performance will be dramatically reduced unless the two levels of processing can be efficiently integrated. An obvious approach is to allow data structures to be shared between the two levels. Having made this general assertion, a practical programming paradigm is required. The framework that we have explored in a number of practical applications involves the idea of a moving focus of attention and will be described below. It is by no means the only possible approach, and indeed can be criticised on the grounds that it involves the use of fairly sweeping heuristics to reduce solution search space. The method has, however, proved useful in automating a number of real visual tasks.

2.2 Focus-of-attention paradigm

In any image-processing system it is, by definition, the role of high-level processing to determine which low-level processing ought to be performed. In a simple system such decisions may be predetermined, but we have already argued that this is unsatisfactory. The focus-of-attention method takes a small step away from complete determinism by allowing the data sets on which low-level processing is performed to be selected dynamically. The simple assumption is made that, although it may be impossible, in a single step, to accurately identify the symbolically significant structures in an image, it is possible to do so approximately. The approximate symbolic description can then be used to identify the image regions that require closer attention in order to obtain a better description. The process can be repeated and provides a controlled method of successively reducing solution search space.

The manner in which a focus-of-attention method can be used to manage the interface between different levels of processing can most easily be explained by briefly describing its application to a real problem. We have chosen to use, as an illustration, a simplified version of the chromosome metaphase analysis software which is part of a routine clinical package developed in this laboratory [8].

An example of a metaphase spread of chromosomes is shown in Fig. 1. The object of the analysis is to recognize individual chromosomes and to label each as one of the 24 unique chromosome types (1–22, X,Y). The main feature used to identify the chromosome type is the banding pattern. Once the chromosomes have been labelled they must be presented in a karyogram, an ordered display sorted by type number (Fig. 2). The simplified analysis sequence is as follows.

1. Get a grey-level histogram of the image and determine a global threshold.

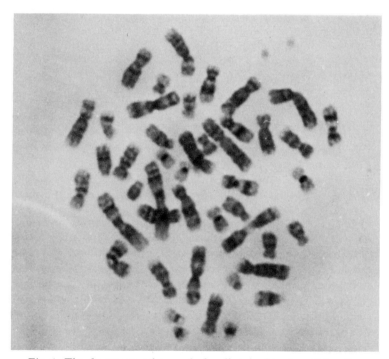

Fig. 1. The chromosomes from a single cell stained to show G-banding.

Integration of symbolic and numerical processing 23

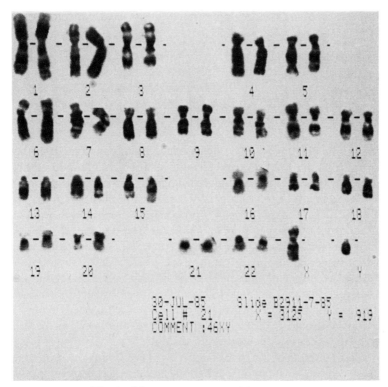

Fig. 2. A karyogram generated from Fig. 1.

2. Threshold the image and find the connected objects in the resulting binary image. These are treated as a rough version of the chromosomes.

3. Get a grey-level histogram of the image in the vicinity of each object and determine a local threshold.

4. Re-threshold the image in the vicinity of each object using the local threshold. Use local connectivity rules to modify each threshold if necessary. If the connectivity rules cannot be satisfied by modifying the threshold in a particular locality then apply a version of the fall set method [9] to segment separate chromosomes.

5. Use curve fitting to obtain a medial axis for each segmented chromosome.

6. For each chromosome, project grey levels from the original image onto the medial axis to obtain an intensity profile. Compare the profile with

a set of standard banding pattern models to select a chromosome type label.

7. Using the type labels, generate a karyogram display. There are pre-assigned slots in the display for two of each chromosome type. Each chromosome is copied into the karyogram image so that its medial axis is vertical. This involves a scale and rotate operation on original image data over the region of the segmented chromosome.

This simple illustration offers many examples of data structures that are involved in both levels of processing. In Step 1 a low-level process generates, from an image, the higher-level histogram structure. High-level processing is involved in syntactically analysing the histogram to obtain a threshold value. In Step 2 low-level processing generates high-level object structures from the image. Step 3 is similar to Step 1 except that the object structures are used to control the low-level process of obtaining a histogram. Step 4 involves an intimate mixture of processing at different levels employing a number of shared data structures. Step 5 is predominantly high-level, and Step 6 once again uses a high-level structure (the medial axis) to control low-level processing. In Step 7 a number of high-level structures generated earlier are used to control the low-level processing involved in generating the karyogram display.

The close coupling between the two levels of processing illustrated by this example suggests that an integrated design methodology ought to be adopted. Particular attention should be paid to the way in which data are stored and manipulated, employing as much shared hardware as possible. This will lead to an architecture where data paths are simplified and overall computational power is significantly enhanced by avoiding unnecessary data transfers.

3 PRACTICAL IMPLEMENTATION

In this section we describe how the ideas outlined above led to the design of two practical image-processing systems which have both been exploited commercially. The Joyce-Loebl Magiscan 2 (M2) was designed and constructed by the authors in 1979/80. The Visual Machines VM1 (also known as CVAS 3000) was designed by the authors in 1983/4 and is architecturally similar. The VM1 offers several enhancements over the M2, but since the differences are not important to this discussion we avoid introducing irrelevant detail and present a description which strictly applies to the M2.

The major goals which we were trying to meet in designing the M2 and VM1 were as follows:

(i) to provide hardware support for the high-level/low-level interface;
(ii) to achieve realistic performance for low-level processing;
(iii) to provide a unified and high-level software development environment;
(iv) to keep within a cost to manufacture ceiling of $20 000.

Although a detailed discussion of these objectives is not appropriate here, the broad aim was to produce self-contained systems, technically and economically suited to undertaking complex visual tasks in the laboratory or factory.

Given these goals, we made the early decision that it was not practical to employ massive parallelism in the system. There has been considerable work showing that goal (ii) can be realized quite effectively using specialized hardware architectures onto which low-level processing tasks map well [1,2,10–15], but we found it difficult to see how goal (i) could easily be achieved by such a system. The dedicated paths that enable efficient low-level processing in cellular arrays constitute a significant problem when communication between low-level and high-level processing is required. We also believed that it would be difficult to achieve goals (iii) and (iv) with a system employing large-scale parallelism.

3.1 Support for high-level processing

The problem of providing hardware support for shared data structures was approached by first proposing an architecture to support high-level processing and then considering how it could be modified to also support low-level processing. PASCAL was chosen as the language in which high-level software would be written, particularly because of its ability to handle complex data structures. The ready availability of transportable compilers and development software was also an important factor, since this significantly reduced the software effort involved in supporting the new architecture. The most straightforward way of supporting PASCAL is to use a compiler generating machine-independent p-code and to design hardware able to emulate a p-machine. This can be achieved using the type of structure shown in Fig. 3. The program and data memory is used to store both p-code and data. During program execution the CPU, controlled by a microprogram, can both interpret p-code instructions and act upon data in memory to generate the appropriate action. The addition of an I/O controller extends such an arrangement into a basic minicomputer.

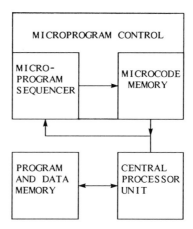

Fig. 3. *A microprogrammed high-level processor.*

The organization described above is used as the basis for high-level processing in the M2. At the time the system was designed it was possible to achieve a microinstruction cycle time of around 150 ns, making conservative use of standard technology. Given a reasonably powerful microinstruction set, this leads to p-code execution times of around 1 μs, which compares very favourably with other implementations of PASCAL [16]. The M2 uses the UCSD PASCAL compiler and development environment. Modifications to this arrangement, which are not relevant to the discussion, allow the VM1 to use a 68000 coprocessor running UNIX with support for both PASCAL and C. We are actively involved in investigating the application of high-level languages which allow more sophisticated symbolic processing to be performed, and hope in the near future to also provide VM1 support for POPLOG.

3.2 Support for low-level processing

Given the high-level processor outlined above, the next problem was to provide low-level processing of adequate performance and to support shared data structures. Previous applications work [8,17–19] suggested that the ability to perform neighbourhood operations on whole images in times of approximately one second would make the use of such methods a practical proposition. A 3 × 3 convolution of a 512 × 512 image in less than one second was adopted as a benchmark for low-level performance, although the same applications experience led us to believe that it would be much too restrictive if this simple class of neighbourhood operation alone were

Integration of symbolic and numerical processing 27

supported. The aim was to provide complete programmability so that arbitrary nonlinear neighbourhood operations could be performed at comparable speed to linear convolution.

The goal of intimately sharing data structures suggested that the CPU, provided for high-level processing, should also be employed for low-level processing. This represented an ideal arrangement since the CPU was already intended to manipulate high-level structures under control of a microprogram emulating the p-machine. The use of the CPU was a practical proposition since, given a suitable instruction set, a multiply and add could be performed in 300 ns (two instruction times). As long as a memory access time of 300 ns could be achieved, a 3×3 convolution involving nine memory reads and one write could in principle be performed on a 512×512 image in 780 ms. To do this it would be necessary to generate image memory addresses with no time penalty and to effect program flow control, again with no time penalty. This was achieved by adding a memory address processor (MAP) and microprogram control processor (MPC) to act in parallel with the CPU. The three processors operate synchronously and are controlled by separate fields in the microinstruction word. The computing power of both the MAP and MPC is comparable to that of the CPU, so that in a tightly coded loop such as the kernel of a neighbourhood operator an effective instruction time of 50 ns is achieved.

3.3 Hardware description

Figure 4 is a block diagram of the complete M2 hardware showing the main interconnection paths. We can now describe each of the main functional elements in some more detail.

3.3.1 Microprogram control

The MPC consists of a $4k \times 48$-bit microprogram memory and a control sequencer. Each microprogram word is divided into three main fields as shown in Fig. 5. The MPC instruction field allows direct or indirect jumps and subroutine calls and subroutine returns all conditional on CPU status flags. Any branch involves only one instruction and incurs no time penalty. The microprogram memory is loaded via the CPU data path normally from a disk file.

3.3.2 Central processor unit

The CPU is a 16-bit arithmetic and logic processor with a 16-bit barrel

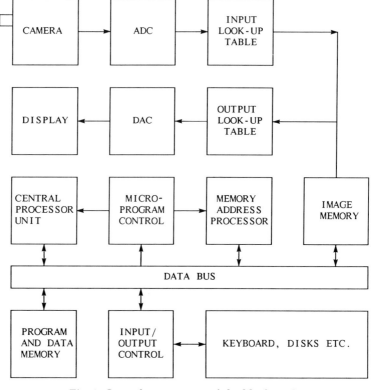

Fig. 4. General arrangement of the Magiscan 2.

shifter, 8 × 8 parallel multiplier, 1024 × 16-bit registers and an accumulator. The general arrangement of the unit is shown in Fig. 6. One operand for the processor is always a register. The same register can also act as the destination. The second operand is one of the other data sources such as the image memory, program memory or MAP. As well as using the selected register as a destination other data destinations such as image memory, program memory or MAP can be addressed.

The barrel shifter appears in the data source input path since its most common use is to extract fields from memory words. In the case of the image memory this allows images of arbitrary word length to be stacked and the data accessed for immediate processing by the CPU. (This also requires bit masking, which is supported in the image memory.) In the case of the program memory the shifter allows efficient field extraction by the p-code interpreter.

Integration of symbolic and numerical processing 29

Fig. 5. Microcode instruction fields.

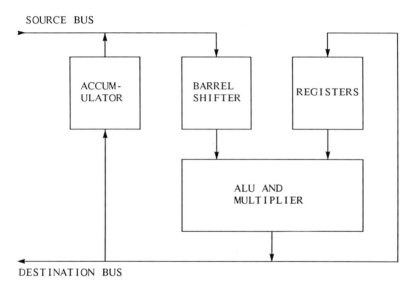

Fig. 6. The central processor unit.

The parallel multiplier allows single-cycle multiplication of operands up to eight bits and is thus suitable for most image data. It operates on the same operands as are presented to the arithmetic and logic processor, and both results are available though normally one will be redundant.

3.3.3 Memory address processor

The MAP is used to compute all access addresses for the image memory. This involves not only microprogram-controlled addressing for the purpose of processing but also memory access for video frame-grabbing and display. The processor also handles light-pen sensing and cursor display and is responsible for refreshing the dynamic memory devices used in the image memory.

The basic arrangement of the MAP is shown in Fig. 7. A 12-bit arithmetic logic processor with sixteen internal registers is used to generate addresses

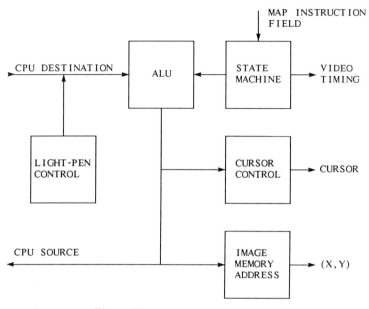

Fig. 7. The memory address processor.

and to carry out other computations necessary to support the light-pen and cursor. A synchronous state machine generates instructions (held in ROM) to control all video related activities and refresh. If an image in the image memory is being displayed, this occupies the processor about 50% of the time. If no stored image is displayed, less than 10% of the time is occupied. Arbitration logic allows microprogram access to the processor whenever it is free. The MAP instruction field allows relative changes in the current X and Y image address pointers, which are held in two registers, and controls the initiation of image memory read or write cycles. Changes in (X, Y) can either be local (± 7 pixels in one instruction) or can use a vector (SX, SY) stored in two other registers to shift between equivalent pixels in two images with different origins. Local changes are used to move around a neighbourhood and through connected regions. The origin shift facility allows for arbitrary allocation of image memory space to source and destination images during image to image transformation.

3.3.4 Image and program memory

There were a number of factors that led to the separation of image memory and program memory. In some ways this was an unfortunate compromise which in any future design should ideally be avoided. The main reasons for

Integration of symbolic and numerical processing 31

separating the two memories were the limited address range afforded by the processor used in high-level processing (CPU, which is a 16-bit machine), the need to provide a direct video input to image memory and the need to support masking on the image memory. Of these the limited address range was the major problem and was virtually insuperable since it was also the case that UCSD p-code only supported a 16-bit address range. Thus, it was impossible to treat images as PASCAL arrays.

The program memory is a 64k × 16-bit memory with 300 ns cycle time and occupies the full address space supported by PASCAL p-code. Since images themselves cannot be held in this memory they are effectively shared between low-level and high-level processing by using a PASCAL record that purports to be the image but is in fact a specification of the image memory space actually occupied by image data.

The image memory is a 1k × 1k × 16-bit memory with a 300 ns cycle time. It is a hardware option to fit a memory with a word size less than 16 bits since this is seldom required by a single image. Images can be of any number of bits (up to 16 bits) and can be stored at any origin in image memory. Images of 512 × 512 or 256 × 256 can be loaded from video or displayed and some support is provided for 1024 × 1024 images. Data masking is provided on input and output, allowing stacked images to be treated independently. Images can be loaded or displayed at any position in memory.

3.3.5 Video input and output

The video input and output paths contain look-up tables which allow arbitrary transformation of the digital video data. The output table is used in the normal way to affect the way in which data are displayed and allows grey-level display, binary overlays, pseudo-colour display and colour overlays. The input table is used for a number of purposes. First it is used to shift image data to the appropriate bit position within the 16-bit memory data word so that images may be directly loaded into any set of planes. Secondly the table is used to perform single-pixel operations such as contrast manipulation or even grey-level thresholding (at multiple thresholds).

3.4 Software description

An integrated software environment is provided for the development of application programs, which involve both high-level and low-level processing.

Microcode programs are generated by a microcode assembler (written in PASCAL) which accepts symbolic program text and generates object files.

Standard microcode exists to implement a PASCAL p-machine on which compiled UCSD PASCAL will execute. Standard microcode is also provided to support a set of primitive manipulations of shared data structures. The facilities provided are sufficiently comprehensive that it is uncommon for even a sophisticated user to require new microcode. The most common reason for generating new microcode is to implement a new type of neighbourhood operator. In this case standard routines provide support for image memory addressing, result scaling, etc. and only the operator kernel need be written.

PASCAL programs are prepared, compiled and linked in the normal way under the UCSD system. Calls to microcode primitives are indistinguishable from calls to PASCAL procedures and the normal user is unaware which of the routines in the library are implemented in PASCAL and which in microcode. The microcode assembler generates link files which are used to statically bind PASCAL and microcode when an application program is link-loaded. When the application program is executed its corresponding microcode file is automatically loaded into microcode memory.

4 DISCUSSION

In the early part of the chapter we presented some general principles which we believe are important to any discussion of image-processing architecture. These ideas influenced the design of the M2 and VM1 systems, but these machines do not, by any stretch of the imagination, fully embody the philosophy.

From a programmer's viewpoint there are two unnecessarily inconvenient features. First is the fact that high-level and low-level code must be written in different languages. The effect of this is significantly reduced by the provision of a comprehensive set of microcode primitives that are sufficiently general that most applications programs can be written without the need to generate new microcode. When new microcode is required the fact that common data structures can be manipulated is important, but the arbitrary distinction between programming methods for the two levels of processing is contrary to the spirit of the system. Similar comments can be made about the arbitrary split between the image memory and the program memory. This creates the additional task for the programmer of managing the allocation of image space. This problem is partially overcome in the VM1 by a software image memory manager, but this is still not an ideal arrangement since it falls outside the normal rules of memory management supported by PASCAL for all other data structures.

Another shortcoming of both the M2 and VM1 is the use of PASCAL (or

C) for high-level processing. Procedural languages are not ideally suited to symbolic processing of any complexity and restrict the choice of analysis strategy. Logic programming or data flow languages would probably be more appropriate, although a multiparadigm programming environment such as LOOPS [20] may be ideal.

The focus-of-attention method is the main framework that we have used in applications programming. As we have already pointed out, this only represents a limited improvement over a completely predetermined solution. The main problem is that the method is basically procedural and, as we have already suggested in the context of languages, this is not entirely desirable. The choice of method was, of course influenced by the languages available. The result of the strictly procedural approach is that it is difficult to generate solutions that are adaptive to the image data that is presented, either at run-time or beforehand during some form of training. In fact the situation is not quite as bad as it seems, since it is possible to embed data driven methods into such a framework. For industrial inspection we have, for instance, used a model matching technique to identify individual mechanical components. We control the application of model matching using the focus-of-attention method. A possible generalization of the method that does not involve the same limitations would use visual cues (not necessarily approximately correct symbolic descriptions) to direct the application of model matching, the results of which could be used recursively as cues.

Finally, the systems as described offer significant but limited computational power for both high-level and low-level processing. In common with most workers in the field, we are interested in the use of VLSI technology to achieve high computational power at reasonable cost. Clearly the use of regular parallelism is becoming a more and more attractive way of achieving this. We strongly believe, however, that it is important to adhere to the basic principles that we have discussed earlier in the paper. SIMD machines do not seem to offer a convenient mechanism for handling the interface between different levels of programming. We still approach architecture from the standpoint that the ability to perform efficient low-level processing should be embedded into a powerful high-level processor.

In summary, practical experience of tackling image-processing applications has led us to develop complementary hardware and software that allow high-level and low-level processing to be integrated in an efficient manner. We know there is still a long way to go!

REFERENCES

[1] Duff, M. J. B. (1979). Parallel processors for digital image processing. In *Advances in Digital Image Processing* (ed. P. Stucki), pp. 265–276. Plenum Press, New York.
[2] Hunt, D. J. (1981). The ICL DAP and its application to image processing. In *Languages and Architectures for Image Processing* (ed. M. J. B. Duff and S. Levialdi), pp. 275–282. Academic Press, London.
[3] Shippey, G., Bayley, R., Farrow, S., Lutz, R. and Rutovitz, D. (1980). A fast interval processor (FIP) for cervical pre-screening. *Anal. Quant. Cytol.* **3**, 9–16.
[4] Ackley, D. H., Hinton, G. E. and Sejnowski, T. J. (1985). A learning algorithm for Boltzmann Machines. *Cognitive Sci.* **9**, 147–169.
[5] Feldman, J. A. and Ballard, D. H. (1982). Connectionist models and their properties. *Cognitive Sci.* **6**, 205–254.
[6] Marr, D. (1982). *Vision*. Freeman, New York.
[7] Mayhew, J. E. W. and Frisby, J. P. (1978). Texture discrimination and Fourier analysis in human vision. *Nature* **275**, 438–439.
[8] Graham, J. and Taylor, C. J. (1980). Automated chromosome analysis using the Magiscan image analyser. *Anal. Quant. Cytol.* **2**, 237–242.
[9] Rutovitz, D. (1978). Expanding picture components to natural density boundaries by propagation methods. The notion of fall set and fall distance. In *Proc. 4th IJCPR, Kyoto, Japan*, pp. 657–664.
[10] Batcher, K. E. (1980). Design of a massively parallel processor. *IEEE Trans. Comp.* **29**, 836–840.
[11] Kruse, B. (1976). The PICAP picture processing laboratory. In *Proc. 3rd IJCPR, Coronado, California*, pp. 875–881.
[12] Uhr, L. and Douglass, R. (1979). A parallel-serial recognition cone system for perception. *Patt. Recog.* **11**, 29–40.
[13] Granlund, G. H. (1981). GOP: a fast and flexible processor for image analysis. In *Languages and Architectures for Image Processing* (ed. M. J. B. Duff and S. Levialdi), pp. 179–188. Academic Press, London.
[14] Gerritsen, F.A. and Monhemius, R.D. (1981). Evaluation of the Delft Image Processor DIP-1. In *Languages and Architectures for Image Processing* (ed. M. J. B. Duff and S. Levialdi), pp. 189–203. Academic Press, London.
[15] Gerritsen, F. A. (1983). A comparison of the CLIP4, DAP and MPP processor-array implementations. In *Computing Structures for Image Processing* (ed. M. J. B. Duff), pp. 15–30. Academic Press, London.
[16] Gilbreath, J. and Gilbreath, G. (1983). Erastothenes revisited: once more through the sieve. *Byte* **8**, 283–326.
[17] Dixon, R. N. and Taylor, C. J. (1979). Automated asbestos fibre counting. In *Machine Aided Image Analysis*, pp. 178–185. IOP Conf. Ser. No. 44, Institute of Physics, Bristol.
[18] Pycock, D. and Taylor, C. J. (1980). Use of the Magiscan image analyser in automated uterine cancer cytology. *Anal. Quant. Cytol.* **2**, 195–201.
[19] Brunt, J. N. H., Taylor, C. J., Dixon R. N. and Gregory P. J. (1983). Theory and practice of applying image analysis to angiography. In *Physical Techniques in Cardiological Imaging* (ed. M. D. Short et al.), pp. 153–162. Adam Hilger, Bristol.
[20] Bobrow, D. G. and Stefik, M. (1981). The LOOPS manual. *Tech. Rep.* DK-VLSI-81-13, *Knowledge Systems Area, Xerox Palo Alto Research Center*.

Chapter Three

Shape Recognition by Human-Like Trial-and-Error Random Processes

M. Nagao

1 REFLECTIONS ON PREVIOUS IMAGE-PROCESSING ALGORITHMS

Research into image processing has made great strides in the past 20 years, and it is now fairly easy to design recognition procedures even for complex objects. However, in the author's opinion the following problems remain to be solved in present image processing technology.

1.1 From rigid algorithms to flexible ones

When analysing the object image the processing algorithms must, of course, be written as a detailed program. There is little that leads us to believe that human recognition processes involve analysis and reasoning as rigid and detailed as computer programs. It seems, rather, that human recognition involves random trial-and-error processes to a great extent. Human recognition of pattern features is not so accurate, but still as a whole gets surprisingly reliable final recognition results. Therefore it seems necessary to incorporate a function that allows the system's peripheral feature detectors to seek the pattern features and feed the answer back autonomously when activated by a higher-level instruction in the computer program. Feature detectors based on this type of function, which we call half-autonomous feature detectors, will greatly clarify and simplify image-processing software, so that we can write pattern-recognition algorithms very easily without going into the detailed description of the low-level processing.

1.2 From hardware realization of low-level processing to that of high-level processing

Efforts are being made to increase processing speed by developing special image-processing hardware, since the work involved in image processing is tremendous. But the hardware developed up to now has practically all been for low-level uniform two-dimensional processing such as noise elimination differentiation and thresholding. A disadvantage here is that these processes occupy an extremely small part of the overall recognition process, and this expensive hardware section does not contribute to overall recognition in the same ratio as the cost of hardware. Much complex work is necessary by present-day recognition systems when attemping to improve the recognition process or to add new recognition objects. The breakthrough in these difficulties will be to couple the low-level processing dynamically with high-level structural descriptions of standard objects to be recognized, and to incorporate a top-down process which drives the low-level processing and gets back the feature parameters supported by hardware.

1.3 Needs for a higher-level shape-description language

In programming, separation of the algorithm and data sections strengthens the toughness of an overall program by allowing changes and additions to the data section. This type of programming methodology should be introduced into image-processing and pattern-recognition processes. This method will enable recognition of multiple subjects within a single system by adding standard object descriptions without modifying the interpretation process and recognition algorithm.

A general outline of a system currently being developed to solve these three problems is given in this paper. A recognition system that can cope with comparatively simple binary object shapes is planned as the first step. The recognition process will be extended to combine binary images including occlusion. The future directions include expansion of the number of recognizable objects and recognition of grey density images.

2 FUNDAMENTAL FEATURE-EXTRACTION MECHANISM

2.1 Flexible-slit method

As is widely known from computer tomography, much can be learned about

Shape recognition by trial and error

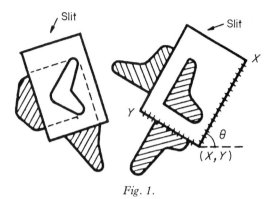

Fig. 1.

the actual shape of an object by examining projection curves of the object. Here we place a rectangular slit at an arbitrary position on a binary graphic image as shown in Fig. 1, and project the graphic image in both the X- and Y-directions. The slit position (x, y), slit direction (θ), slit size and slit spacing are to be freely settable. Various predictions can be made about the object by analysing the projection curves in the X- and Y-directions from this slit. Several slit applications can be performed semirandomly to obtain significant projection curves which can suggest the existence of special shape features. The new position of the slit is calculated from the information of the past projection curves. The projection curves are also analysed to confirm whether the expected shape is in fact correct. Recognition of the projection curves obtained from the slit is therefore necessary for many typical image shapes.

2.2 Feature extraction by the slit method

There are the following three objectives for the analysis of the projection curves:

(i) estimation of a better location for a new slit in relation to the object by a rough classification of the projection curves;
(ii) feature extraction and pattern recognition of the projection curves;
(iii) correlation of a projection curve with a reference curve to identify the class of the curve.

Process (i) enables the consistent extraction of features in (ii) and (iii) by positioning the slit for optimum extraction. Correlation with a standard projection curve in (iii) is the most direct method for discrimination of complex curves that cannot be identified by (ii). One advantage here is that

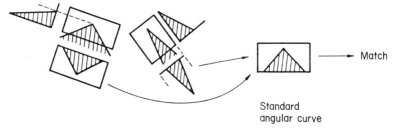

Fig. 2.

various different curves can be identified by using a single standard curve because the projection curves can be adjusted by varying the slit parameters and by setting the slit in a more appropriate position. For example, if we apply a slit in an appropriate position, angle and sample slit spacings for the different angular shapes shown in Fig. 2, the projected curve obtained will be approximately the same, and matching can be carried out using a single standard curve. After getting a good match for an input angle with a standard angular curve, the actual angle size and orientation of the input angular shapes can be calculated from slit parameters such as sample spacing and direction.

The problem here is how to place the slit in an appropriate place. The slit is first placed at random at the position where the shape is assumed to be. The projection curve obtained here is analysed by method (i), as will be explained next in detail, to give estimated parameters which allow the slit to be repositioned and the orientation and size changed appropriately. The best matching of the projection curve is obtained by repeating this trial-and-error process three or four times until the optimum slit position is found.

Let us look at an example of detecting a sharp angle. As shown in Fig. 3, the first slit is placed arbitrarily. From the projection curves of the slit, rotation through a certain angle and a new slit size is suggested. This process is repeated several times to finally obtain the best position and size for the angle, which is symmetrical about the centre axis of the slit as shown in Fig. 4. This position is called the "normal position".

When the pair of projection curves is as shown in Fig. 5, the two-dimensional image is estimated as shown in Fig. 6. Then the next slit should be extended in the Y-direction, and also extended in the X-direction to get a better slit position as shown in Fig. 7.

The normal position for a circle, rod end, line edge and rod body are shown in Fig. 8.

If the projection curve and the standard curve cannot be satisfactorily matched regardless of the number of times the slit position is changed, the

Shape recognition by trial and error

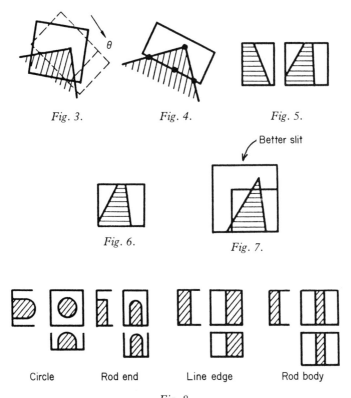

Fig. 3. Fig. 4. Fig. 5.

Fig. 6. Fig. 7.

Circle Rod end Line edge Rod body

Fig. 8.

object shape is considered to be different from the predicted one (or a shape that cannot be detected by this method), and the test is regarded as unsuccessful.

2.3 Feature parameters of a projection curve

The feature parameters are to be calculated for projection curves in case (ii) to predict the general shape of the projection curve. The following parameters are calculated.

Parameters to be calculated for a projection curve (see Fig. 9):
slit size (w steps × h steps);
    h_{10} maximum value of projection curve;
    h_i $10 \times i\%$ of h_{10};

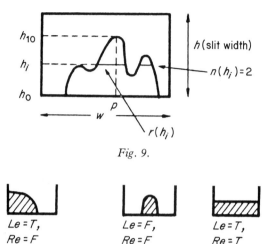

Fig. 9.

Fig. 10.

$r(h_i)$ maximum width of curve at height h_i;
$n(h_i)$ number of crossing on the curve at height h_i;
Le, Re logical value for the curve determining whether it touches the left side or right side (see Fig. 10).

Parameters obtained by hardware:
(1) h_{10} and its position P;
(2) whether $h_{10} = h$ or not (Cd);
(3) area of the projection curve (S);
(4) $r(h_i)$ and $n(h_i)$ for $i = 0, 1, \ldots, 9$;
(5) Le and Re;
(6) $n(h_o)$.

Parameters calculated by microprogram:
(7) mean of heights $h = S/r(h_o)$;
(8) variance of heights σ_h^2;
(9) mean of curve positions $\bar{x}$;
(10) variance of curve positions σ_x^2;
(11) minimum height in the projection curve.

Parameters from the differentiated projection curve:
(12) position of the maximum;
(13) positions of local maxima.

Shape recognition by trial and error 41

2.4 Trial-and-error process for feature extraction by the slit method

A variety of shapes can be estimated to exist from the parameters introduced in Section 2.3. For example, when the orientation of the slit A in Fig. 11 is approximately in line with the longer side of the L-shaped rod as shown in the figure, the variance of the projection height values will be a near-minimum.

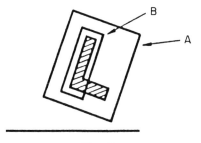

Fig. 11.

Therefore, when it is detected that the projection width at 80% of the height is almost the same as the width at 60%, a rod shape may be predicted. It is also possible to predict the orientation, width and length of the predicted rod. If we place slit B as shown in Fig. 11 from the calculated parameters of the slit projection curves of A, it just covers the rod exactly. Recognition of the rod shape is thus confirmed.

When placing the slit based on accurate information such as with slit B in Fig. 11, an analysis can be made with fairly high reliability. This is an advanced stage of analysis. When there is little information about the shape of the projection, there is no alternative to placing the slit at random. As the projection curve obtained from the slit will contain various uncertainty factors due to this randomness, it becomes pointless at this stage to make accurate calculation of complex components such as the projection-curve parameters.

The first step necessary is to determine where the next slit should be placed from the information obtained by the parameters of the projection curves of the previous slit. We have the following results.

Range of existence of projection curves

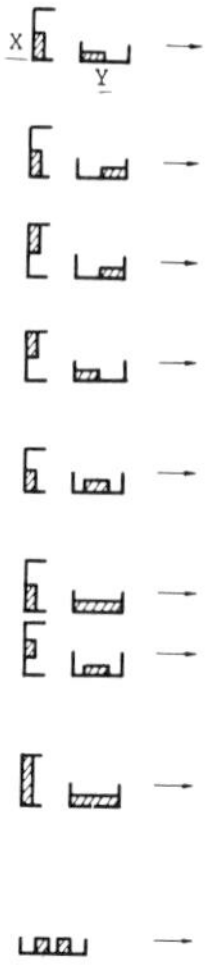

x↑ ▭ —	rotate −45°, extend the width of Y;
▭ ▭ —	rotate +45°, extend the width of Y;
▭ ▭ —	rotate +45°, extend the width of Y;
▭ ▭ —	rotate −45°, extend the width of Y;
▭ ▭ —	rotate so that the position of Y_{max} comes to the centre of the projection curve;
▭ ▭ —	extend the width of Y;
▭ ▭ —	isolated object—rotate so that σ^2 for the projection curve becomes a minimum;
▭ ▭ —	if the height of the curve is relatively small, rotate randomly to see if a different situation occurs;
▭ —	divide into two slits.

We have to check the possibilities of the combinations of the projection curves shown in Fig. 12. Physically possible combinations of these as X- and Y-projection curves are surprisingly small. Meaningful combinations are those shown in Fig. 13. The figures are two-dimensional shapes, which produce the projection curves written under each shape.

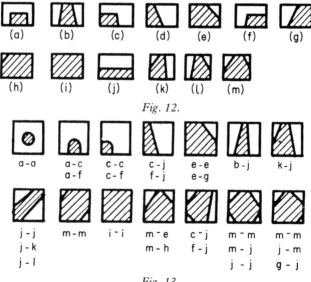

Fig. 12.

Fig. 13.

Shape recognition by trial and error 43

For all these curve combinations we can estimate the positions and the size of the next slit. Next actions for the slit placement are the following, for example:

$\left.\begin{array}{l}\text{c--c}\\\text{c--f}\end{array}\right\}$ rotate;

$\left.\begin{array}{l}\text{c--j}\\\text{f--j}\end{array}\right\}$ get vertical edge; extend in the edge direction;

$\left.\begin{array}{l}\text{e--e}\\\text{e--g}\end{array}\right\}$ rotate and extend in the edge direction;

b--j   extend in the rod direction;

$\left.\begin{array}{l}\text{k--j}\\\text{j--j}\\\text{j--k}\\\text{j--l}\end{array}\right\}$ rotate and get b--j projection combination; extend in the rod direction;

$\left.\begin{array}{l}\text{m--m}\\\text{i--i}\\\text{c--j}\\\text{j--j}\\\text{etc.}\end{array}\right\}$ enlarge the slit;

Most typical projection curves will be obtained by repeating this process. Typical ones are the angle, rod end, half-circle, rod body and line edge as shown already in Fig. 8.

3 TWO-DIMENSIONAL PATTERN-MATCHING FUNCTION

Although various methods have been described here for identification of the projection curves of the object shapes in the X- and Y-directions through a slit, there is a limit to the recognition of complex shapes with these methods based on the projection curves. Many objects are difficult to describe using simple projection-curve information. The two-dimensional input image will have to be compared with a two-dimensional standard pattern for recognition by two-dimensional matching. The method of two-dimensional matching is divided into two categories, one for isolated single objects, and the other for complex objects that are composed of two or more partial figures, each of which is to be matched two-dimensionally with a standard pattern.

3.1 Pattern matching of a single isolated shape

The method of two-dimensional matching when a single isolated object to be recognized is at an arbitrary position is as follows.

(1) Place the slit of a size that adequately covers the single object in any orientation at random, and obtain the X-, Y-projection curves. Using the parameters of these curves, determine the principal axis of inertia. The next slit is placed along this axis of inertia. An alternative method is to find out the longest direction of the object by applying the slit several times by rotating the direction. Then the next placement of a slit is along this axis in the longest direction.

(2) Place the slit parallel to the principal axis of inertia and adjust the slit so that the object is centred in the slit. Centring in the slit means a slit width of $1.1x$, $1.1y$, when the dimensions of the object in the X- and Y-directions are assumed to be x, and y.

The processes in (1) and (2) are illustrated in Fig. 14.

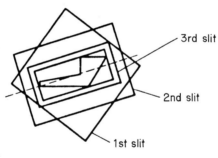

Fig. 14.

(3) Construction of a two-dimensional standard template is exactly the same as (1) and (2); that is, the recognition of an object shape. One example is the third slit in Fig. 14. Irregularities in the input shape should be absorbed to a certain extent in this instance, and a "don't care" zone is introduced. A two-dimensional standard pattern obtained in this way is shown in Fig. 15. Although the best matching position is not sharply obtained by this matching method with "don't care" zones, the characteristics are very effective against noise components at the edges, and for the matching of a part of a complex shape.

(4) Matching of the object and the standard pattern obtained in this

Shape recognition by trial and error

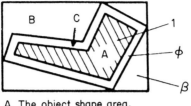

A The object shape area, which has the value 1;
B The background area;
C Don't care zone

$$\underset{A}{\Sigma 1} = \underset{B}{\Sigma \beta}, \underset{C}{\Sigma \phi} \simeq 0.1 \underset{A}{\Sigma 1}$$

Fig. 15.

manner is carried out at the third slit position in Fig. 14, and the best match is obtained in this third slit after shifting laterally a certain number of degrees from the principal axis of inertia.

3.2 Partial pattern matching in complex shapes

This two-dimensional pattern matching can also be used for partial pattern matching as follows.

(1) If a large slit as in Fig. 16 is placed at various angles, it may be predicted from the projection curves that the object has a rod in the R_1 direction.

(2) A long narrow slit (slit 1) is then placed at this position, and slit 2 is determined by the length of the rod from the Y-projection of slit 1.

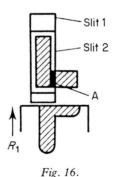

Fig. 16.

(3) Two-dimensional matching is then carried out by the pattern of slit 2 against a standard rod pattern. Although area A in the diagram tends to interfere with the matching function, the effect will be minimal because it falls in the "don't care" area.

(4) Next, if Steps (1), (2) and (3) are carried out for the remaining area after temporarily masking the already identified parts, the remaining area can be identified as a short rod. In this way the angle formed by the two rods can be obtained. Although the second slit is used as an erase mask when carrying out recognition of the remaining area, if this erase mask is used in Step (1) only, and is not used in the remaining stages, the recognition will become as shown in Fig. 17.

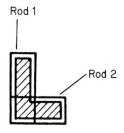

Fig. 17.

There is another possibility for a recognition process in cases such as the shape placed as shown in Fig. 18, and the primary features detected as shown by circles. That is, from the output curve of the slit of Fig. 18, the extreme corners can be detected easily as the crossing points of the end points a, b, a′, b′ of the projection curves. Then at these points the angle-detection routine is activated and the angle directions, angle values and some other information are obtained. Then the connecting edge lines are obtained by the edge-detection routine activated by the information of the angle edge directions. In this way the contour figure of the object can be obtained with

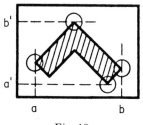

Fig. 18.

Shape recognition by trial and error 47

the new findings of the two corners that were not detected at first. Therefore the object description must be not only a component subregion description, but also include contour and some other descriptions. Even by the subregion description, a complex object can be decomposed into different subregion combinations, and all these must be accepted by the recognition system. This is very important for the recognition of occluded objects.

4 IMAGE-DESCRIPTION SYSTEM

4.1 Requirement for an image-description system

Although image analysis and recognition are carried out using fundamental feature-extraction functions as explained here, the best approach is one in which a structural description of the object is prepared and in which the object description and the feature-extraction function are coupled in an interpretive routine which carries out overall recognition. Since this interpretive routine will be a general process, it must be capable of operating even if a new object description is given. Other functions that it should incorporate are (i) a hierarchical structural description of shapes; (ii) capability of holding multiple descriptions of the same graphic portion; (iii) trial-and-error processes included as part of the interpretation routine and a function allowing decisions to be made for only those parts of the given input figure that can be confirmed from the image descriptions; (iv) the function to decide by relative comparisons which object, out of a group of objects, is to be recognized, by using the smallest number of obtained features, rather than proceeding with overall indiscriminate recognition.

Generally speaking, recognition of shapes should not be too rigorous or strict. The general features should be extracted and the interpretation made within that range. If there is no match with the final target, recognition should proceed by finding other detailed features. If two input objects cannot be separated from the description and interpretation of a figure, object descriptions must be enriched; that is much more detailed descriptions must be given which enable discrimination of the two objects. Such descriptions must be given in hierarchical order from global to local and be exact for the interpretation system to be able to go top-down step by step to get the recognition result by the earliest stage of analysis.

4.2 Shape-description system structure

A tentative description system of shapes is as follows:

(1) object;
(2) name of component parts;
(3) description of the relationship of component parts;
(4) output parameters of the object.

An example of the description of an L-structure such as in Fig. 17 is as follows:

(1) L-STRUCTURE
(2) A, B; rod
(3) A. DIR⊥B. DIR
    A. END ≃ B. END

Another description of the same L-structure is

L-STRUCTURE
SIZE = max (A. LENGTH, B. LENGTH)
WIDTH = min (A. LENGTH, B. LENGTH)

In the first description we have two items that are partial components of the L-structure, denoted as A and B; this line indicates that the objects A and B are "rod". The next line describes the relation between A and B. It shows that A. DIR (direction of A) and B. DIR are at right angles to each other and that A. END (end position of A) is at the same position as B. END. Matching of angles and position will always be approximate. In other words, a tolerance of $\pm 10\%$ for angles, and $\pm W$ (W is width of rod) for position will be permitted. A. END and B. END both have 2 values and the system checks for a match between any of the 4 combinations.

An example of a complex object is given in the following (see Figs 19 and 20).

Flypan   Cubepipe
Fig. 19.   *Fig. 20.*

Shape recognition by trial and error

Flypan

A: rod
B: circle
A. END + [B.R*cos(A.DIR),B.R*sin(A.DIR)]
  ≈ B.CENTER

Cubepipe

A: rod
B: circle
A.END + [B.R*cos(A.DIR + 90),0] ≈ B.CENTER

5 SYSTEM IMPLEMENTATION

Hardware and software are being developed to incorporate the above algorithm in a single system. A general outline of this system is shown in Fig. 21.

Preprocessing, such as removal of noise, is carried out by the general image-processing hardware (B) at the speed of memory cycles per pixel of the image memory (A). The image is then properly thresholded and sent to binary memory (C). The data in (C) and sometimes in (A) are sent to (D), where the processings explained earlier are executed by hardware. As the slit is replaced a number of times in the slit method, the memory readout time makes up a large portion of the processing time. Algorithms for placing optional size slits in optional positions and of getting values from grid points

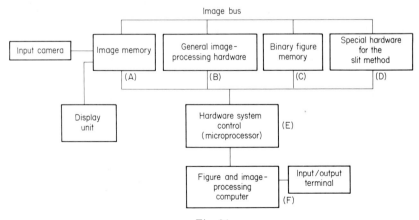

Fig. 21.

of a slit are all built into the hardware and are contolled by commands from (E).

The major part of the processes explained in Section 2 is supported by hardware, including two-dimensional correlation. Complex processes such as the calculation of variances are computed by the microprocessor in (D). For example, the time required to obtain the main results from a single positioning of a 32×32 slit is $32 \times 32 \times 0.2$ µs $+ 32 \times 0.1$ µs $+ \alpha = 210$ µs (approx.) if the cycle time of the binary memory is taken as 0.2 µs. All calculations such as rotation of coordinate axes and density values by linear approximation at required grid points are processed by hardware within one cycle of binary memory. This means that one feature can be fetched in about 2 ms even if the slit is replaced 10 times to get it. Although this is not exceptionally fast, we believe this is sufficient compensation when we consider the advantage of not having to write a program in detail to fetch each of the features.

6 CONCLUSION

This system is programmed to recognize comparatively simple two-dimensional shapes existing in isolation. Although the use of this system for recognition of objects which overlap each other to a certain extent will be a subject for the future, we believe that this will not be an excessively difficult problem. This is because it will simply be a question of detecting the features of partial figures and registering that these features possibly represent separate objects. A certain amount of additional study will no doubt be necessary in expanding this system to the recognition of variable-density images, and this will hinge on how well the general image-processing section (B) in the block diagram can construct the binary figure. Although a binary image memory is used in this system mainly because of the problem of cost, we believe that the same system concept can be used for the analysis of variable-density images with the same basic slit method by modifying memory (C) to store grey density images, and also by modifying the hardware (D) for multivalue data processing.

GENERAL REFERENCE

Nagao, M. (1984). Control strategies in pattern analysis. *Patt. Recog.* **17**, 45–56.

PARALLEL ALGORITHMS

Given that architectures for low-level processing may not be, and probably cannot be, optimum for intermediate-level processing, and given that a range of potentially effective intermediate-level architectures can be proposed, there remains the problem of selecting or designing algorithms that map well onto the new architectures and that take advantage of the mode of parallelism they offer. As yet, no clearly defined rules of procedure have been developed to assist in this process.

Leah Jamieson introduces the topic and extends it to take into account the choice of a programming language. She shows how it is possible to tabulate and relate key characteristics of algorithms, architectures and languages in an attempt to find a good match between them. Hungwen Li and his colleagues describe, in the next chapter, their experimental language "V", which they have developed as a research tool for benchmarking architectures and algorithms, this language enabling them to express a parallel architecture and a parallel algorithm in the same language. In the following chapter, Peter Gemmar studies some specific practical problems concerned with the analysis of aerial photographs and shows how the parallelism inherent in the actual task can be exploited in a carefully optimized parallel processing system. Finally in this second section, Robert Hummel carries out an in-depth and extremely complex analysis of the Shiloach/Vishkin connected component algorithm, using it as a vehicle to explore the relative performance of SIMD and MIMD structures implementing the same algorithm.

Chapter Four
The Mapping of Parallel Algorithms to Reconfigurable Parallel Architectures

L. H. Jamieson

1 INTRODUCTION

In both parallel architectures and parallel algorithms there exist many design choices for which there are no direct counterparts in conventional serial processing. In architectures, examples of such choices include number of processors, processor–memory configuration and interconnection topology. In parallel algorithms, issues that do not arise in serial programming include determination of the number of processors needed/useful for a task, data allocation across memories, and interprocessor communication requirements. The move to parallelism has introduced new degrees of freedom to both the architecture and algorithm design process. For effective use of parallel systems, it is essential to obtain a good match between algorithm requirements and architecture capabilities.

In this chapter, parameters for characterizing the execution characteristics of parallel algorithms are presented. The specific context of this work is in the design of an intelligent operating system for a large-scale reconfigurable parallel architecture so that, given a new algorithm and a specification of performance requirements, the operating system can select an architecture configuration and a mapping of the algorithm to the configuration such that the required performance is achieved.* The

---

* Throughout this chapter, we will refer to performance requirements without specifying the performance criteria being used. This in itself is a complex question (see e.g. [1]). A reasonable initial assumption is that speed is of utmost importance; more involved criteria may be appropriate in specific environments.

approach is also applicable to the design of algorithmically specialized parallel computers, where a set of algorithms, characterized by the union of their attributes, dictates design decisions for the parallel architecture. The focus of this paper is in defining a framework in which parallel algorithms can be discussed and in identifying characteristics of parallel algorithms that have the greatest effect on their execution.

2 THE VIRTUAL-ALGORITHM MODEL

Although there is no formal notion of the scope of an algorithm in terms of how "complicated" a problem a single algorithm solves, it will be convenient to assume that an *algorithm* performs a single well-defined function, and that a *task* is performed by execution of a collection of algorithms. From an applications point of view, the objective will always be to perform a task, with the selection/design of algorithms and architectures being the means by which the task is accomplished. Thus a task might be "identify an object in a noisy image," with component algorithms including filtering, edge detection, contour tracing, scale and orientation normalization, computation of a Fourier descriptor, and template matching. Clearly there are many issues relating to the interaction of the algorithms that comprise a task; however, our focus here will be on the individual algorithm.

In order to establish at what point an algorithm is in some sense "bound" to an architecture, the "algorithm development life cycle" shown in Fig. 1 is assumed. Using the area of digital signal processing for the purpose of illustrating the components of the graph, an example of a *problem statement* might be "obtain a spectral estimation of a given signal." An *algorithm approach* might be the decision to use an FFT to obtain the spectral estimation, without specifying any details of the FFT implementation. The *virtual algorithm* represents the point at which the computational steps to be performed have been determined, but the mapping of those steps to an architecture has not yet been committed. In the signal processing example this might correspond to the decision to use a radix-2 decimation-in-time formulation of the FFT. The virtual algorithm might be represented in terms of a dataflow or signal flow graph. The *ideal algorithm* will correspond to the parallel algorithm that would be selected if no constraints were placed on the architecture configuration. In general, this will be the best known parallel implementation of the virtual algorithm. In the example, for computation of an N-point FFT and assuming speed as the performance criterion, this might be an $N/2$-processor, $\log_2 N$-step SIMD algorithm using a perfect shuffle or cube-type interconnection network [2–4]. The ideal algorithm may in fact never be used, but may suggest strategies for

Parallel algorithms and architectures 55

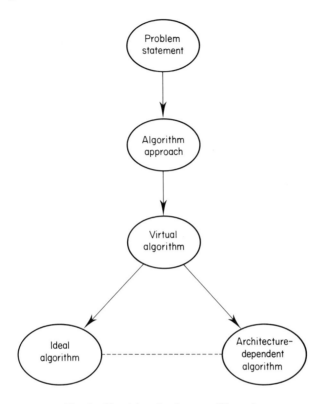

Fig. 1. *Algorithm development life cycle.*

implementation of the algorithm and provides a reference point against which architecture-dependent algorithms can be compared. This potential (but possibly indirect) relationship between the ideal algorithm and architecture-dependent algorithms is indicated by the dashed arc in Fig. 1. The *architecture-dependent algorithm* represents the binding of the algorithm steps to a particular architecture. In the signal-processing example this would include decisions about the number of processors, data allocation and interprocessor transfers to be used.

In order to solve the "task-to-architecture" mapping problem in its entirety, all five points on the time line must be addressed. Thus, in the context of performing a task, a number of different algorithm sets may be possible, corresponding to a number of distinct *problem statements* for the possible component algorithms. Given a *problem statement*, the selection of an *algorithm approach* and *virtual algorithm* will necessarily depend largely on the specific application area. A true representation of the *task* solution space replicates the algorithm development life cycle for each algorithm in a number of candidate sets of algorithms (representing the alternative approaches to performing the task), but also allows branching at the point of choosing the *algorithm approach* (e.g. alternative algorithms for obtaining a spectral representation) and at the point of specifying the *virtual algorithm* (e.g. alternative FFT formulations).

In this chapter we focus our attention on the transition from a *virtual algorithm* to an *architecture-dependent algorithm*. At the point of specification of the virtual algorithm, the computational steps to be performed have been determined, but the "structure" of the implementation—the allocation of data, the assignment of computational steps to processors—has not yet been defined. It is the mapping from the virtual algorithm to the architecture-dependent algorithm that will determine the actual execution profile. Using the virtual algorithm as our starting point, we identify characteristics of algorithms and architectures that bear on the expected performance of architecture-dependent algorithms.

3 ARCHITECTURE MODEL

We will assume a target parallel architecture with the general attributes listed below. The purpose of the assumptions is to define a very general architecture framework, so that no undesirable restrictions are imposed on the architecture by the assumptions. Rather, by examination of the algorithm attributes, the subset of capabilities required in the architecture can be selected. Thus the framework is more general than most existing parallel systems, but many architectures, including PASM [5], Ultracomputer [6], the Cosmic Cube [7], the Butterfly [8], the Connection Machine [9], MPP [10], Clip 4 [11] and pyramid architectures [12,13], can be embedded in the framework. The principal classes of architectures not covered by the assumptions are dataflow systems and heterogeneous architectures.

Assumptions for the target architecture are as follows.

(i) The system consists of a large number of homogeneous processors. In mapping the virtual algorithm to an ideal parallel architecture, the number of processors available is considered to be unlimited. The resulting algorithm will be the ideal algorithm for the class of architectures being considered.

(ii) The system can be organized with processors accessing a shared global memory or with each processor having an associated local memory.

Parallel algorithms and architectures   57

(iii) The system is partitionable into independent submachines of varying sizes. This implies that for each algorithm the partition size can be selected to meet the needs of the algorithm. (For consideration of a single algorithm, this assumption merely states that the machine size can be specified. In the more general case of considering a set of algorithms comprising a task, the partitioning assumption means that a number of algorithms can be executing simultaneously, on different partitions of the system.) The partitioning can be changed dynamically during execution.

(iv) Each partition of the system (and the entire system itself) is capable of both SIMD and MIMD operation, and can dynamically switch between modes during execution.

(v) The system has a flexible interconnection network which can provide a wide variety of communications patterns within each partition.

Given these assumptions, the problem becomes one of selecting the architecture configuration—memory organization, partition size, mode of operation, network configuration—and then mapping the virtual algorithm of that configuration.

4 ALGORITHM CHARACTERISTICS

Parallel algorithms can be characterized in many ways, along a number of dimensions (see e.g. [14–16]). Similarly, parallel architectures can be described in terms of a number of attributes (see e.g. [17–19]). Some relations between image processing tasks and parallel architectures have been discussed by Cantoni and Levialdi [20] and by Aggarwal and Yalamanchili [22]; Chiang and Fu [21] have proposed a graph model for relating algorithm and architecture attributes. Ideally, a set of orthogonal characteristics would describe parallel algorithms, and a corresponding set of orthogonal characteristics would describe parallel architectures, with a unique bijection performing the mapping from one to the other. Since this is clearly not the case, a less formal approach is taken as a first step in expressing the relation between algorithms and architectures. Table 1 lists a set of characteristics that influence the execution of a virtual algorithm on a reconfigurable parallel architecture; Table 2 lists one possible set of attributes by which an architecture can be parameterized. Focusing on the algorithm characteristics, informal descriptions of some of the dependencies are traced.

Table 1
Algorithm characteristics

Nature of the parallelism:
  data parallelism vs. functional parallelism
  data granularity
  module granularity
Degree of parallelism
Uniformity of operations
Synchronization requirements
Static/dynamic character of the algorithm
Data dependencies
Fundamental operations
Data types and precision
Data structures

Table 2
Architecture characteristics

Number of processors
Processor–memory organization
Homogeneity
SIMD/MIMD/pipeline
Synchronization mechanisms
Connectivity
Granularity
Memory size
Word length
Addressing modes
Data types
Data structures supported
Processor capability
I/O

(i) *Nature of the parallelism.* Under this heading come a number of attributes having to do with the "kind" of parallelism that is used and the way in which the algorithm and/or data can be decomposed.

(a) *Data parallelism vs. functional parallelism.* Parallelism can be achieved by dividing the data among the processors, by decomposing the algorithm into segments which can be assigned to different processors, or by macropipelining. The

Parallel algorithms and architectures

type of parallelism will affect the allocation of data, the assignment of processes to processors, and the basic decision as to what mode of parallelism (SIMD/MIMD/macropipeline) to use.

(b) *Data granularity.* Data granularity will have to do with the "size" of the data items processed as a fundamental unit, and will have a bearing on the data allocation, communications requirements, processor capability and memory requirements.

(c) *Module granularity.* Module granularity [14] will have to do with the amount of processing which can be done independently of other processes. It is essentially a measure of the frequency of synchronization, and will affect the choice of SIMD vs. MIMD operation, the assignment of processes to processors, the communications requirements, and the likelihood of equalizing the execution times of component parts of the algorithm.

(ii) *Degree of parallelism.* This will be related to both the data granularity and the module granularity. Its most direct impact will be on the choice of machine size and on the maximum speedup attainable.

(iii) *Uniformity of the operations.* If the operations to be performed are uniform (e.g. across the data or feature set), then SIMD (or pipeline) processing may be feasible. If the operations are not uniform, then MIMD processing will be chosen and strategies to equalize the computational load across the processors may come into play. These strategies may be applied statically at compile time or dynamically at execution time.

(iv) *Synchronization requirements.* In addition to the synchronization requirements implied by the process granularity, consideration of precedence constraints is implicit in characterizing the synchronization requirements. This will affect the assignment of processes to processors and the scheduling of various components of the algorithm.

(v) *Static/dynamic character of the algorithm.* The pattern of process generation and termination will affect the processor utilization, the scheduling of subprocesses, the memory organization and the communications requirements.

(vi) *Data dependencies.* The data dependencies in an algorithm will play the largest role in dictating data allocation patterns and communications characteristics. They will also have a major part in the decision regarding global vs. local memory organization. Extensive work on characterizing communications with respect to a number of parallel architectures has been reported by Levitan [23].

(vii) *Fundamental operations.* The basic operations performed in the algorithm will dictate the processor capabilities needed.

(viii) *Data types and precision.* The atomic data types and data precision will bear most directly on the individual processor capability and on the memory requirements, but may also imply requirements for communications bandwidth.

(ix) *Data structures.* Many algorithms can be characterized as having a "natural" data structure (or structures) on which operations are performed. The ability of an architecture to support the needed access patterns, to exploit possible regularity in the structures, and to allow the needed interactions between parts of the structures will affect algorithm performance.

5 LANGUAGE CONSIDERATIONS

The virtual-algorithm model facilitates isolation of various stages of the algorithm development process, and in particular allows us to focus on the specific mappings from the virtual algorithm to the ideal and architecture-dependent algorithms. One issue not addressed is the representation of the algorithms. Because the virtual algorithm consists of what is in a sense an abstract specification of computational steps, graph representations—dataflow graphs, signal flow graphs, operator nets, Petri nets—appear to provide a natural representation of the virtual algorithm. Many aspects of the mapping from the virtual algorithm to an architecture-dependent algorithm (e.g. degree of parallelism, synchronization requirements, static/dynamic character of the algorithm, data dependencies) can be formulated as comparisons between attributes of the graph and parameters of the architecture. Although the graph representation is appropriate for studying the algorithm–architecture relationships, in practice—i.e. when the algorithm is actually implemented and executed—a specific language will be used to represent the algorithm. The language interacts with both the algorithm and the architecture: for good performance, the constructs and structures used in the algorithm should be supported by the language, and the language constructs used should be implemented efficiently in the architecture. The performance of the algorithm will therefore ultimately depend on the

Table 3
Language characteristics

Concurrency specification
Synchronization specification
Mechanisms for information sharing:
  global variables
  messages
Control flow operations
Function and procedure calling mechanisms
Data types and precision
Data structures
Addressing modes
Primitive operations
I/O

goodness of the three-way match: algorithm, architecture and language. Table 3 gives some of the language characteristics that will enter into the matching process. The need to consider the language as a factor in performance has implications at both the very early system-design (and possibly language-design) stages and at the time of formulating the virtual and architecture-dependent algorithms.

6 SUMMARY

Because of the interplay of algorithm, architecture and language, the problem of designing efficient parallel algorithms is a complex one. Our focus here has been on the algorithm: defining a framework in which parallel algorithms can be discussed and identifying features that will characterize the execution of parallel algorithms. Concentrating on the relationship between the virtual algorithm and architecture-dependent algorithms has led to a set of algorithm features which can be related to the choice of an architecture configuration. Clearly it is possible to change "point of view," e.g. to ask "given a parallel architecture, for what types of algorithms is it best suited?" or "what parallel architecture features are required to support efficient execution of a given language?" The feature-based approach outlined here, although informal, should have the power to be able to deal symmetrically with the myriad permutations of the questions that can be asked about how parallel algorithms, architectures, and languages fit together.

ACKNOWLEDGMENTS

Discussions with many colleagues are gratefully acknowledged: George Adams, Ed Delp, Bob Douglass, Art Feather, Dennis Gannon, H. J. Siegel, Phil Swain, Len Uhr and Bob Voigt. This research was supported by the United States Army Research Office, Department of the Army, under Grant DAAG29-82-K-0101, by the United States Air Force Command, Rome Air Development Center, under Contract F30602-83-K-0119, and by SRI.

REFERENCES

[1] Siegel, L. J., Siegel, H. J. and Swain, P. H. (1982). Performance measures for evaluating algorithms for SIMD machines. *IEEE Trans. Software Engrg* **8**, 319–331.
[2] Stone, H. S. (1971). Parallel processing with the perfect shuffle. *IEEE Trans. Comp.* **20**, 153–161.
[3] Pease, M. C. (1977). The indirect binary n-cube microprocessor array. *IEEE Trans. Comp.* **26**, 458–473.
[4] Siegel, L. J., Mueller, P. T., Jr., and Siegel, H. J. (1979). FFT algorithms for SIMD machines. In *Proc. 17th Ann. Allerton Conf. on Communications, Control, Computing*, October, pp. 1006–1015.
[5] Siegel, H. J., Schwederski, T., Davis, N. J., IV, and Kuehn, J. T. (1984). PASM: A reconfigurable parallel processing system for image analysis. In *Proc. Workshop on Algorithm-Guided Parallel Architectures for Automatic Target Recognition*. July, pp. 263–291. (Reprinted in *ACM SIGARCH Computer Architecture News*, **12** (Sept. 1984), pp. 7–19.)
[6] Gottlieb, A., Grishman, R., Kruskal, C. P., McAuliffe, K. P., Rudolph, L. and Snir, M. (1983). The NYU Ultracomputer—designing an MIMD shared memory parallel computer. *IEEE Trans. Comp.* **32**, 175–189.
[7] Seitz, C. L. (1985). The Cosmic Cube. *Commun. ACM* **28**, 22–33.
[8] Rettberg, R. D. (1984). The Butterfly multiprocessor. In *Proc. Computer Architecture for Vision Workshop*, May, pp. 1–3.
[9] Knight, T., Poggio, T. et al. (1984). Connection machine architecture and vision algorithms. In *Proc. Computer Architecture for Vision Workshop*, May, pp. 31–36.
[10] Batcher, K. (1980). Design of a massively parallel processor. *IEEE Trans. Comp.* **29**, 836–840.
[11] Duff, M. J. B. (1982). Parallel algorithms and their influence on the specification of application problems. In *Multicomputers and Image Processing* (ed. K. Preston and L. Uhr), pp. 261–274. Academic Press, New York.
[12] Tanimoto, S. L. and Klinger, A. (eds.) (1980). *Structured Computer Vision: Machine Perception through Hierarchical Computation Structures*. Academic Press, New York.
[13] Uhr, L. (1983). Pyramid multi-computer structures, and augmented pyramids. In *Computing Structures for Image Processing* (ed. M. J. B. Duff), pp. 95–112. Academic Press, London.

[14] Kung, H. T. (1980). The structure of parallel algorithms. In *Advances in Computers*, Vol. 19 (ed. M. C. Yovits), pp. 65–112. Academic Press, New York.
[15] Swain, P. H., Siegel, H. J. and El-Achkar, J. (1980). Multiprocessor implementation of image pattern recognition: a general approach. In *Proc. 5th Int. Conf. on Pattern Recognition*, December, pp. 309–317.
[16] Etchells, R. D. and Nudd, G. R. (1983). Software metrics for performance analysis of parallel hardware. In *Proc. Image Understanding Workshop*, June, pp. 137–147.
[17] Hockney, R. W. and Jesshope, C. R. (1981). *Parallel Computers: Architecture, Programming and Algorithms*. Adam Hilger, Bristol.
[18] Smith, B. W. and Siegel, H. J. (1985). Models for use in the design of macro-pipelined parallel processors. In *Proc. 12th Ann. Symp. on Computer Architecture*, June, pp. 116–123.
[19] Danielsson, P. E. and Levialdi, S. (1981). Computer architectures for pictorial information systems. *Computer*, **14**, 53–67.
[20] Cantoni, V. and Levialdi, S. (1982). Matching the task to an image processing architecture. In *Proc. 6th Int. Conf. on Pattern Recognition*, October, pp. 254–257.
[21] Chiang, Y. P. and Fu, K. S. (1983). Matching parallel algorithm and architecture. In *Proc. 1983 Int. Conf. on Parallel Processing*, August, pp. 374–380.
[22] Aggarwal, J. K. and Yalamanchili, S. (1984). Algorithm driven architectures for image processing. In *Proc. Workshop on Algorithm-Guided Parallel Architectures for Automatic Target Recognition*, July, pp. 1–31.
[23] Levitan, S. P. (1985). Evaluation criteria for communication structures in parallel architectures. In *Proc. 1985 Int. Conf. on Parallel Processing*, August, pp. 147–154.

Chapter Five

The V-Language for Polymorphic Architectures and Algorithms

H. Li, C.-C. Wang and M. Lavin

1 INTRODUCTION

Computer architectures that are highly replicable from a single processing element (PE) in a regular pattern [1–8] have long been recognized as suitable for image processing. Of great theoretical interest is the question of whether there exists an optimal matching of a range of problems in image processing to the aforementioned architectures. Previous research [9] indicated that one architecture (or interconnection of PEs) can be optimal only for a portion of the problem domain. With the advance of the VLSI technology, the belief that a PE can support polymorphic architecture (or multiple interconnections) and that such a polymorphic arrangement can expand the domain of optimal matching is growing stronger.

A polymorphic array consists of multiple processing elements connected by a framework featuring multiple regular patterns. Each processing element is a microprocessor-like arithmetic logic unit with local storage for code or data and several links to communicate with other processing elements. These links constitute multiple regular interconnection patterns embedded in a polymorphic "parent" architecture. A morphosis (or a "child" architecture) can be dynamically "derived" from the polymorphic parent architecture by instructions. Transformation from one morphosis to another is controlled by a controller, and the communication among processing elements is synchronous. Such polymorphic arrays are in favour of VLSI implementation, and the parent architecture is furnished during the chip fabrication.

Although generic polymorphic arrays that can derive any morphosis (interconnection) are not available at present, there do exist architectures that support more than one interconnection pattern. The MPP [4] has a baseline square-mesh interconnection. However, it can be rearranged, on demand, as a linear array (or pipeline), a set of column arrays or a set of row arrays. A binary n-cube [10] can induce square-mesh connection and accordingly linear arrays. The baseline morphosis of AAP [5] is an octagonal mesh. For speed and fault tolerance reasons it can short-circuit the row connection such that a more efficient morphosis is created for communication. The CHiP [11] is not only hexagonal-mesh-connected but also has an embedded tree. Along the line of polymorphic architecture, research effort has been devoted to the augmented networks [12] such that more communication patterns can be subsumed by a generic polymorphosis.

The polymorphic arrays have a wide spectrum of applications such as image processing, matrix operations, finite-element methods, computational geometry and digital signal processing as demonstrated by the aforementioned architectures and their "monomorphic" counterparts including pipelined [2], array [3], tree [13], pyramid [6–8] and systolic [1] architectures.

To take full advantage of polymorphic arrays, softwares must be designed to match the underlying morphosis appropriately. However, it can be shown that creating a parallel algorithm that closely matches a parallel architecture is a significant task [14]. Having versatile and dynamic communication patterns at the designer's disposal certainly extends the flexibility of the algorithm design and matches the algorithm better with the architecture. This is the motivation for promoting the polymorphic arrays and the V-language.

Existing languages are not sufficient for precisely expressing algorithms to be executed on a polymorphic architecture. The CSP message-passing mechanism [15] adopted by the Cosmic Cube [10] is powerful for asynchronous communication but is not well suited for the synchronous communication of the polymorphic arrays. The Parallel PASCAL [16] language developed for MPP [4] provides generic ways to manipulate array data but lacks the capability of expressing the communication pattern required by algorithm. The lack of morphic expression may be compensated by elaborate compiling schemes to deliver an efficient code for "monomorphic" machines. But for the "polymorphic arrays", it is a nontrivial task for a compiler to generate an efficient code without knowing explicitly the desired morphosis in the algorithm designer's mind.

An interesting approach to explicitly expressing the morphic information is the Poker [17] environment for CHiP [11]. In Poker the description of mapping from processes to processors and the interconnection among processors are separated from the algorithm description. Instead of being expressed in language, the description of morphosis and mapping is

The V-language

specified with the assistance of a graphic package. Such a graphic facility allows the expression of a static communication pattern at process level. When a process requires a dynamic change of morphosis within itself, Poker has to break the process into finer granularity and subscribe another morphosis. Since polymorphic algorithms frequently change communication pattern for efficiency, there is a need to integrate the algorithm and architecture description in the same language. This paper deals with this issue by introducing the concept of "structured process" and many V-constructs which support the "structured process" attribute.

We describe the concept of "structured process" and the V-language in Section 2. In the same section we discuss several communication types frequently found in polymorphic algorithms and show how the "structured-process" attribute can be used to express the interprocess communication more precisely. In Section 3 a sample polymorphic algorithm is presented to crystallize the "structured-process" concept. A discussion of how the algorithm matches the polymorphic arrays follows. In Section 4 we describe an environment consisting of a translator and an emulator for V and the use of the environment in benchmarking parallel architectures and algorithms. Finally, in Section 5 we summarize the work and point out a few directions for future research.

2 STRUCTURED PROCESS AND THE V-LANGUAGE

We took the approach of adopting the C-language as the base language then adding to C five constructs to form the V-language. The five constructs are created to support the "structured process" to be defined.

2.1 Definition of structured process

A *structured process* [18] is a group of tightly coupled processes that share the same process name and are structured according to their pattern. Common patterns used in polymorphic algorithms include the straight line (in which one process has two neighbours in the structure), the square mesh (in which one process has four neighbours), the hexagonal mesh (in which one process has six neighbours), the tree (in which one process has one parent process and few son processes), the binary n-cube (in which one process has n neighbours) and the pyramid (in which one process not only has neighbours at the same level but also has a parent at the upper level and children at the lower level). Figure 1 shows these popular patterns. At the end of Section 2.2 we shall offer the replacement of the graphic patterns in Fig. 1 by their

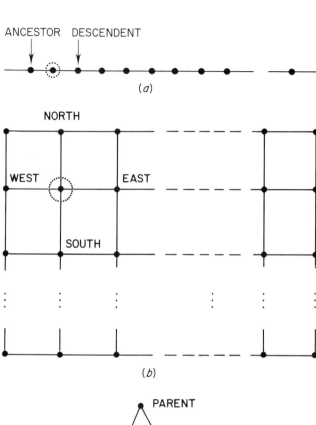

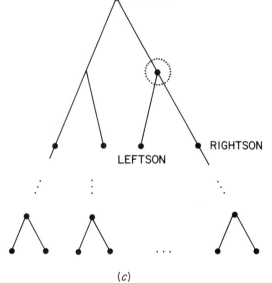

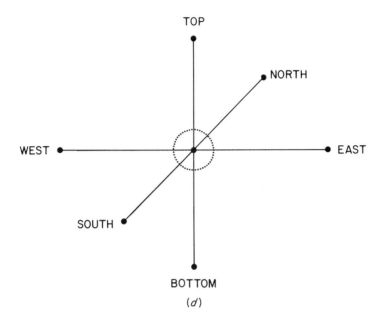

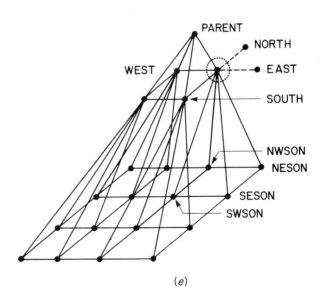

Fig. 1. Popular communication patterns: (a) pipelined and linear systolic architectures; (b) 2D array architecture; (c) binary tree architecture; (d) cube architecture; (e) pyramid architecture.

corresponding descriptions in terms of a structured process. More interesting patterns can be found in [19].

The declaration of a structured process creates a replication of identical or similar processes related by the declared structure. Each member of the structure is a process and is assigned a Process IDentification (PID) derived from the structure. The PID can be a scalar or an M-tuple, where M is the dimension of the declared structure. For processes that follow the aforementioned patterns, the PIDs have a simple and regular relation which is isotropic. The isotropic property is common to polymorphic algorithms and can be utilized to express the interprocess communication more precisely. This will be elaborated further in Sections 2.2 and 2.3. We mentioned that processes within a structure are identical or almost identical except for minor differences. The discrepancy among processes in a structure is usually due to the difference in PID caused, for example, by mapping a large array to different processes. To eliminate the discrepancy, we treat PID as a condition and use the condition to unify processes. The advantage of this approach is that only one version of code is needed for all processes in the same structure and highly symmetrical "SEND/RECEIVE" rendezvous can usually be detected, leading to an efficient utilization of the underlying morphosis. This phenomenon is demonstrated by the example in Section 3. In summary, the PID of the structured process is treated as an "attribute" which can be interrogated in the polymorphic algorithms and used as an explicit linkage between the algorithm and architecture. This viewpoint will be further elaborated in Section 3.

To show how the PIDs are derived from the structure, we declare a two-dimensional structured process by the statement

structured process hist[16][16](parameter list).

The above declaration creates 256 (16 × 16) processes. All of them share the same process name "hist" and each of them carries a parameter list as typical procedures (e.g. PASCAL). These processes are distinguished from each other by their PIDs ranging from [0][0] to [15][15]. So far, and without further definition of a pattern, the relation among the processes in the structure is missing and we are not able to use the structure and the PIDs to accomplish efficient interprocess communication. We define the pattern in Section 2.2.

2.2 Patterns in structure

A "pattern" in a structure is declared with the following syntax:

#strdef < pattern-name > [(< parameter-list >)]
(< exp >) ... (< exp >)

The V-language

where #**strdef** is the keyword to signal the declaration. It is followed by a pattern name. A pattern name is an identifier which can be used in the algorithm to facilitate interprocess communication. Following the pattern name is a list of optional parameters which specify a dynamic pattern frequently encountered in polymorphic algorithms. Usually, these parameters consist of loop indices used in the algorithms. One or more expressions are mandatory after the parameter list. The total number of expressions is equal to the dimensionality of the structure. More precisely, the first expression specifies the PID of the source/destination process in the first dimension of the structure and the second expression for the second dimension, etc. Operands of these expressions include the parameters specified in the parameter list, the PIDs and other constants. The modular N rule (where N is the maximum PID value in any dimension) applies to the evaluation of an expression which guarantees that the resultant value is in the range from 0 to N. This allows a "toroidal" communication pattern or a "folding" between processes with the smallest and the largest PID values in any dimension.

For example, we can define the two-dimensional square-mesh pattern as

#**strdef west (PID0)(PID1 − 1)**
#**strdef east (PID0)(PID1 + 1)**
#**strdef north (PID0 − 1)(PID1)**
#**strdef south (PID0 + 1)(PID1)**

In such a structure, each process has four neighbours: the west, the east, the north and the south neighbour, each of them designated by a PID-tuple < **PID0,PID1** >, where **PIDk** is the PID value of the kth dimension. Note that the PID expression is independent of the physical "location" of the "reference" process. In other words, the pattern is isotropic. Being isotropic, the definition of a pattern in a structure is independent of the choice of a coordinate system. This is different from the Poker environment, which depends on a grid system. Given a morphosis or a physical interconnection pattern, a drawing of the coordinate system as provided by Poker will help the algorithm design. For the structured process, such a drawing is however not necessary.

The above pattern is *"static"* in the sense that the pattern remains unchanged for the life of the structure. Being "static", it does not require the optional "parameter list" which is to define a "dynamic" pattern. In the following paragraph, an example of a dynamic pattern is shown.

The statement

#**strdef east(i) (PID0)(PID1 + i∗2 + 1)**

defines the dynamic east neighbour of a process in the square mesh. By changing the parameter **i**, a process may use the same expression east(**i**) to communicate with many processes bearing a regular pattern. For example, east(**i**) designates the process with **PID = (PID0)(PID1 + 1)** when **i = 0** and the process with **PID = (PID0)(PID1 + 3)** when **i = 1**. In fact, the example defines a "tree" pattern with parameter **i** equivalent to the level of the tree. This happens for example when data are initially distributed in a linear array and the minimum or the maximum of the data are required at a certain stage of the algorithm. The tree pattern allows the algorithm to be executed faster than in a linear-array pattern.

We also allow a *"compound"* pattern in a structure by defining the new compound pattern based on existing ones as shown below:

#strdef < pattern-name > { < pattern-name >
... < pattern-name > }

The pair of curly brackets contains the existing pattern names and distinguishes them syntactically from the new pattern name. For example, the collection of the "west", "east", "north" and "south" neighbours in the previous example can be represented by a new pattern name "neighbours" defined by

#strdef neighbours {west, east, north, south}.

With the hierarchical nature of the compound pattern, it is possible to formulate an "aggregate" which is useful for broadcasting communication. More than one level of the hierarchy is possible. However, a two-level hierarchy is the most useful.

In retrospect, we are able to replace the common patterns for the polymorphic processes in Fig. 1 by the description in terms of the structure and the pattern definition. They are shown in Fig. 2.

2.3 Interprocess communication in structure

Based on the structured-process concept and the pattern, six types of interprocess communication can be defined to facilitate the description of the polymorphic algorithms. These types are self, relative (static and dynamic), relative group (static and dynamic) and absolute. Each of them will be described in the following paragraphs.

```
#strdef ancestor (PID0-1)
#strdef descendent (PID0+1)
```

Figure 2a. Pipelined and Linear Systolic Architectures

```
#strdef east  (PID0)(PID1+1)
#strdef west  (PID0)(PID1-1)
#strdef south (PID0+1)(PID1)
#strdef north (PID0-1)(PID1)
```

Figure 2b. 2-D Array Architecture

```
#strdef parent  (PID0-1)(PID1>>1)
#strdef leftson (PID0+1)(PID1*2)
#strdef rightson (PID0+1)(PID1*2+1)
```

Figure 2c. Binary Tree Architecture

```
#strdef top    (PID0)(PID1)(PID2+1)
#strdef east   (PID0)(PID1+1)(PID2)
#strdef west   (PID0)(PID1-1)(PID2)
#strdef south  (PID0+1)(PID1)(PID2)
#strdef north  (PID0-1)(PID1)(PID2)
#strdef bottom (PID0)(PID1)(PID2-1)
```

Figure 2d. Cube Architecture

```
#strdef parent (PID0>>1)(PID1>>1)(PID2-1)
#strdef east   (PID0)(PID1+1)(PID2)
#strdef west   (PID0)(PID1-1)(PID2)
#strdef south  (PID0+1)(PID1)(PID2)
#strdef north  (PID0-1)(PID1)(PID2)
#strdef nwson  (PID0<<1)(PID1<<1)(PID2+1)
#strdef neson  (PID0<<1)((PID1<<1)+1)(PID2+1)
#strdef swson  ((PID0<<1)+1)(PID1<<1)(PID2+1)
#strdef seson  ((PID0<<1)+1)(PID1<<1)+1)(PID2+1)
```

Figure 2e. Pyramid Architecture

Fig. 2. Description of architectures as structured processes.

A general communication type can be expressed as

var2 = var1.pattern; or
var2.pattern = var1;

The former is for RECeiving and the latter for SENDing. Evaluation of the pattern results in one or more PID-tuples which identify the owner of **var1** in REC case and that of **var2** in SEND case. In addition to the aforementioned static, dynamic and compound patterns, we also allow the pattern to be "null' or a "PID-tuple" to express the "self" and the "absolute" communication type respectively.

The *self communication type* assumes a null pattern (e.g. **var2 = var1;**) which indicates that **var1** resides in the same process as **var2**. No interprocess communication occurs. It is clear for the algorithm designer, the reader and the compiler to understand that all variables in the statement are accessible by the very process without going through the morphosis.

The *relative type* is used for the communication between one sender and one receiver specified by a non-compound pattern which can be static or dynamic. Using the same example of square mesh, the relative static communication can be expressed as

var2 = var1.west;

while the relative dynamic communication is expressed as

var2 = var1.west(j);

where the "west" designates a process whose PID maintains a known "relative" position to the reference process.

This "relative" dispersion is a very important piece of information to the polymorphic arrays. With this information, a morphosis can be selected to achieve the most efficient communication; the links in the morphosis can be scheduled; and if **var1** and **var2** are arrays, the indices are the same for all processes in the structure because of the isotropic property. Consequently, the code generation can be analysed according to the property of the "relativeness".

The *relative group type* is based on the compound pattern. It invokes communication among more than two processes in the structure. An example of the relative group type is to broadcast **var1** to four neighbours (e.g. **var2.neighbours = var1;**), where "neighbours" are defined as before. Both static and dynamic cases are possible for the relative group communication type. Note that the relative group SENDing is a one-to-many communication, while its RECeiving counterpart (e.g. **var2 = var1.neighbours;**) is a many-to-many communication hence requiring **var2** to be expanded to a vector of the size of "neighbours".

The V-language

Similar to the relative type, the known relative dispersion is informative in programming the polymorphic arrays. Furthermore, it is worth noting that the expression power is not limited to the "port" concept as suggested by Poker [17] and the Cosmic Cube [10]. The "neighbours" defined by the group communication need not have a direct "port" between the source and destination processes. This feature significantly increases the expression power for a polymorphic algorithm and eliminates the restriction to a specific hardware.

The *absolute communication type* employs a constant PID-tuple (e.g. [2,4]) in place of a pattern. It is always static. This communication type is similar to the message passing mechanism of CSP [15] except that the PID is explicitly known in the structured process while the PID in CSP is assigned by operating system. Because of this subtle difference, each PE of the polymorphic array does not need a local operation system (or routing table) to perform the "message" routing. This feature leads to a significant reduction of the PE complexity and largely increases the number of PEs per equivalent VLSI chip area. This feature is not only true for the absolute communication type but also for all communication types.

2.4 Parameter binding and activation of a structured process

In the above paragraphs, declaration of a structured process is defined. Execution of a structured process, however, needs to be explicitly activated. Furthermore, formal parameters declared in the structured process must be bound to actual parameters (like procedures and functions) before invocation. Parameter binding is usually implied and combined with procedure/function calls for most high-level languages. In V we decided to separate the parameter binding from the structured-process invocation for the reason that the expression for parameter binding in a structured process is usually lengthy (as will be seen in the following paragraphs). Separation of parameter binding from invocation leads to semantic clarity and readability of programs.

Binding actual parameters with formal parameters of a structured process is accomplished by the **map** statement with the following syntax:

map < parameter > [< index list >] of < structure name >
= < variable > < exp list >

The keyword **map** is used to inform the compiler that the statement is for parameter binding. It is followed by a parameter name declared in the structure-process statement. When the parameter is an array, an index list is used to specify the range of the indices of all dimensions. (A scalar can be mapped with an index list.) The structure name is the structure to which the parameter belongs. The variable name to the right of the equal sign is an array name which serves as the actual parameter (or argument). The **exp list** (or expression list) indicates the relation between the actual parameter and the formal parameter. An example of **map** is shown as follows:

map limage[i:1 .. 5][j:2 .. 6] of hist = gimage [PID0 + i][PID1 + j]

The above statement binds the formal parameter **limage[i][j]** of process **hist[PID0][PID1]** with the actual parameter **gimage[PID0 + i] [PID1 + j]**, where the ranges of **i** and **j** are 1 to 5 and 2 to 6 respectively. For each formal parameter declared in the "structured process" statement (see Section 2.1), there is at least one **map** statement to initialize the parameter.

Following the **map** statement, an **activate** statement is used to invoke a structured process. An **activate** statement has the following syntax:

activate < process list >

To activate the **hist** process defined in Section 2.1, for example, the following statement is used:

activate hist[0 .. 15][0 .. 15]

It is possible to activate more than one structured process in the same **activate** statement. For instance,

activate hist[0 .. 15][0 .. 15], average[0 .. 1023][0 .. 1023]

Semantically, 16×16 **hist** processes are invoked simultaneously with 1024×1024 **average** processes.

With the above definitions, we can proceed to illustrate the application of the structured process. In the following section, the problem of calculating the histogram of an image is programmed in V using the concept of a structured process.

3 USING THE STRUCTURED PROCESS

In this section we use an example shown in Fig. 3 to demonstrate how the structured process and its related features can be used to design a polymorphic algorithm. We then show how the algorithm can be better mapped to a polymorphic array with the explicit PID as a linkage.

3.1 Designing a polymorphic algorithm

The example is to compute the histogram of an $N \times N$ image on an $M \times M$

The V-language

```
00    structured process hist[M][M](limage)
01    int limage[N/M][N/M];
02    {
03    #strdef east(piddyn)  (PID0)(PID1+(1<<piddyn))
04    #strdef west(piddyn)  (PID0)(PID1-(1<<piddyn))
05    #strdef south(piddyn) (PID0+(1<<piddyn))(PID1)
06    #strdef north(piddyn) (PID0-(1<<piddyn))(PID1)
07    int h[256], i, j; /* h stores histogram */
08    int mask1, mask2;
09         /* stage 1 computes the histogram of subimage */
10         for(i=0; i<N/M; i++)
11         for(j=0; j<N/M; j++)
12              ++h[limage[i][j]];
13
14         /* stage 2 computes histogram per row */
15         for(j=0; (1<<j)<M; j++) {
16              mask1=~(~0<<j);
17              mask2=~(~0<<(j+1));
18              if(((PID1&mask1)==mask1)&&((PID1&mask2)==mask2)) {
19                   for(i=0; i<256; i++)
20                        h[i] += h[i].west(j);
21              }
22              else if(((PID1&mask1)==mask1)&&((PID1&mask2)!=mask2)){
23                   for(i=0; i<256; i++)
24                        h[i].east(j)=h[i];
25              }
26         }
27
28         /* stage 3 computes histogram on column */
29         for(j=0; (1<<j)<M; j++) {
30              mask1=~(~0<<j);
31              mask2=~(~0<<(j+1));
32              if(((PID0&mask1)==mask1)&&((PID0&mask2)==mask2)
33                                         &&(PID1==(M-1))) {
34                   for(i=0; i<256; i++)
35                        h[i] += h[i].south(j);
36              }
37              else if(((PID0&mask1)==mask1)&&((PID0&mask2)!=mask2)
38                                         &&(PID1==(M-1))) {
39                   for(i=0; i<256; i++)
40                        h[i].north(j)=h[i];
41              }
42         }
43    }
```

Fig. 3. The histogram algorithm.

square-mesh architecture with enhanced morphosis. The image is decomposed into $M \times M$ subimages, each of which has size $(N/M) \times (N/M)$ (assuming that N is a multiple of M) and is distributed (or mapped) to the processing element with the corresponding PID. The algorithm is expressed in C and the extended constructs of V.

The algorithm is constructed in three stages. At stage 1 (lines 9 to 12) each processing element calculates the histogram of its assigned subimage and stores the partial result in array **h**. Since all data are local, no interprocess communication is involved. At stage 2 (lines 14 to 26) the PEs in the same row cooperatively formulate the partial histogram of that row; the partial histogram is stored in the rightmost column (highest **PID1**). To accomplish this in $O(\log M)$ iterations, a "tree" pattern or a "logical pyramid" is required. Two conditions **mask1** and **mask2** therefore were created out of the PIDs such that the morphosis is uniquely prescribed by the algorithm. Related to the **mask1** and **mask2** conditions, the **if** and **else if** statements (lines 18 and 22) eliminate the difference of the processes such that one unified version of code is sufficient for all processes in the structure. We will discuss how to use the structure as a condition shortly.

Finally at stage 3 (lines 28 to 43), the PEs in the rightmost column finish the histogram computation cooperatively and the histogram is stored in the element with PID-tuple **[M-1][M-1]**. Since only one column of PEs are active, we created two other conditions **mask1** and **mask2** to unambiguously impose that the other columns stay idle during stage 3. These two conditions also specify a tree pattern and by using them the code is uniform as discussed previously.

Using a structure as a condition. At both stages 2 and 3, the algorithm is executed for $\log M$ iterations in which not all PEs are active (i.e. some processing elements are disabled). The condition **mask1** in lines 16 and 30 is created for each element to test whether it should be active (enabled). At the 0-th iteration of stage 2 all PEs are active, while at iteration 1 the PEs with odd PID1 are active and those with even PID1 are disabled. At iteration 2, PEs having PID1 with "all one" pattern for the last two bits (e.g. 3,7, . . etc.) are active and others are disabled. The enabled/disabled state at the subsequent iterations is reflected by the bit pattern of **mask1** accordingly.

The other condition expressed by **mask2** is to control the "SEND" or "REC" (receive) action for proper communication. At the 0-th iteration of stage 2, the processing elements with even PID1 SEND while the elements with odd PID1 RECeive. The SEND/REC action of the subsequent iterations can be understood by deriving the bit pattern of **mask2**.

Using the structure as a condition to control the activeness of the process allows us to design and document the algorithm precisely. In the sample histogram algorithm, the activity of each process is exactly prescribed without ambiguity. More specifically, among the three possible states: idle, SEND and RECeive, each process is associated with one and only one state for every combination of PID0, PID1 and the loop index **j**. This is a significant improvement over the conventional English-like algorithm description with the aid of an interconnection and data flow graph as has been seen in many systolic algorithms [20].

The V-language

High clarity is also contributed by intentionally discarding all "patchy" activities which occur when the operations of the cooperating processes are not identical but the underlying architecture is synchronous. In such situations, different patches are added to each process type such that the operations of all processes are eventually in concert and fit the architecture. Because the patches are "artificial" instead of required by the original algorithm, the result of patching is always a confusing and misleading algorithm. Using the structure as a condition accomplishes the same uniformity but avoids patches.

The **if** and **else if** statements (lines 18 and 22) use the structure as a condition to avoid patches and to accomplish uniformity. The actions of the **if** and **else if** statements are a pair of SEND/RECeive (lines 24 and 20). They form a synchronous rendezvous for interprocess communication according to the structure declared at the beginning of the process (lines 3 to 6). Although there are three process types (idle, SEND and REC) at stage 2, we need only one version of code for all processes.

Prescribing morphosis. As mentioned earlier, the histogram computing can be accomplished in $O(\log M)$ iterations via a "tree" or "logical pyramid" pattern. Such a wish is strongly suggested by the algorithm designer in the structure declaration and pattern definition (line 0 and lines 3 to 6) as well as in the **for** statements (lines 15 and 29). Having a parameter **piddyn** in the pattern definition, the east, west, south and north neighbours are dynamic and the desired dynamic morphosis is prescribed by the loop index **j**. According to **j**, the desired morphosis at the 0-th iteration of stage 2 is an $M \times M$ square mesh because every PE is active and communicates with its immediate east or west neighbour. (In fact, the south and north paths in the mesh are not even used.) Then at iteration 1 of stage 2, the desired morphosis is an $M \times (M/2)$ rectangular mesh. The desired morphosis can be equivalently derived from the observation that the distance between the east/west neighbours is two units (i.e. **piddyn** = **j** = 1). For the subsequent iterations, the morphosis follows a shrinkage factor of 2 in the row direction until it reaches the size of $M \times 2$.

At stage 3 the desired morphosis is similar to that of stage 2, except that only the rightmost column is of interest. Therefore, the morphosis starts from an $M \times 1$ to a 2×1 array at a shrinkage factor of 2.

The structured process offers the programmer a facility to prescribe the desired morphosis within the algorithm. Such a capability is different from Poker in that the structure process can be integrated in the algorithm while the Poker maintains separate descriptions for architecture and algorithm. Another feature of the structured process is the capability to prescribe a dynamic morphosis within a process.

3.2 Mapping algorithm to polymorphic array

The desired dynamic morphosis for the histogram algorithm is a mesh augmented with a "2D tree" or "logical pyramid" pattern. It can be embedded in a square mesh which is used for computing the histogram of the subimages. After computing the local histogram, the partial answer is stored in **h**-arrays distributed in $M \times M$ PEs in the mesh; the final answer is the summation of these **h**-arrays. We wish to show how the desired morphosis can be embedded in the square mesh.

At 0-th iteration of stage 2, the east/west neighbours use the links of the mesh to communicate. As described earlier, at iteration 1 every other column is skipped. Equivalently, we can design the PE to detect its own PID condition so that when it is "disabled" the PE will "short circuit" its west and east links. The "short-circuit" action leads to a tree pattern along the row direction. For stage 3, similar "short-circuit" activity in the column direction will allow the histogram algorithm to be mapped optimally to the desired pattern.

The implementation of such a polymorphic array based on known popular patterns (such as a square mesh) is a low-risk approach. No extra path needs to be added to the baseline pattern. Furthermore, the "short-circuit" capability has been built in several processors [6] but was used for other purposes. Besides these known techniques, a polymorphic array does require the PEs to recognize its PID condition.

4 THE V-ENVIRONMENT

To validate the correctness and evaluate the efficiency of the polymorphic architectures and algorithms, an environment consisting of a translator and an emulator for V has been developed as a research tool for investigating the interaction between polymorphic architectures and algorithms. As the input to the tool, the user specifies both the algorithm and architecture in V and obtains, as the output, the functional result for correctness validation and the execution statistic for efficiency evaluation.

The translator translates programs written in V into a C program and thus saves development effort. The implementation of the translator is based on a parser generator. The V-grammar is added to the C-grammar and then fed into the parser generator to produce a parser for V. On top of the parser, the semantic rules were written for each V-construct to constitute the translator. This approach takes advantage of the existing C-environment and does not need a significant amount of effort in developing a V-compiler.

The emulator accepts the C-program after translation, and reproduces the parallelism prescribed in the original V-program. Within the kernel of the emulator is a monitor which manages the execution flow. When a structured process is encountered, the monitor "parallelizes" the structure and proceeds the execution of each process within the structure until the point where a synchronization is needed. Means to extract the statistics (such as execution time and waiting time) are also part of the monitor.

5 CONCLUSION

We have developed an experimental language V based on the concept of "structured process" through which a group of intimately coupled processes can be known to each other via publicizing their PID-tuple. Such an attribute is especially useful for a class of architectures called polymorphic arrays which have multiple regular interconnection patterns and are suitable for image processing.

With a structure, a polymorphic algorithm can be precisely designed and documented. High clarity and readability are accomplished by structured processes integrated within the algorithm. Using the structure as a condition, the code for similar processes can be unified such that only one version of the code is necessary. The algorithm also reveals a strong suggestion for a desired optimal morphosis through the declared structure and the PID. We believe that the structure and PID are informative enough for a compiler to produce efficient code for the polymorphic arrays.

To fully test the concept of a structured process and investigate the behaviour of a polymorphic architecture and algorithm, we have developed an environment for V. The environment consists of a translator from V to the C-programming language and an emulator of V to functionally mimic the parallelism specified in the language. Such an environment not only allows us to benchmark the performance of the polymorphic architecture/algorithm, but also enables us to evaluate the relative merit of two morphoses. With such a research tool, the optimal matching between architecture and algorithm for a range of problems in image processing and computer vision is being studied.

REFERENCES

[1] Kung, H.T. (1982). Why systolic architectures? *Computer* **15**, 37–46.
[2] Sternberg, S. R. (1983). Biomedical image processing. *Computer* **16**, 22–34.
[3] Duff, M. J. B. (1976). CLIP 4: a large scale integrated circuit array parallel processor. In *Proc. 3rd Int. Joint Conf. on Pattern Recognition*, pp. 728–733.

[4] Batcher, K. E. (1980). Design of a massively parallel processor. *IEEE Trans. Comp.* **29**, 836–840.
[5] Kondo, T., Nakashima, T., Aoki, M. and Sudo, T. (1983). An LSI adaptive array processor. *IEEE J. Solid-State Circ.* **18**, 147–156.
[6] Dyer, C. R. (1982). Pyramid algorithms and machines. In *Multicomputers and Image Processing* (ed. K. Preston and L. Uhr), pp. 409–420. Academic Press, New York.
[7] Tanimoto, S. L. (1981). Towards hierarchical cellular logic; design considerations for pyramid machines. *Dept. Comp. Sci. Tech. Rep.* 81-02-01, *Univ. Washington.*
[8] Uhr, L. (1972). Layered "recognition cone" networks that preprocess, classify, and describe. *IEEE Trans. Comp.* **21**, 758–768.
[9] Aggarwal, A. (1984). A comparative study of x-tree, pyramid and related machines. In *Proc. 25th Symp. on Foundations of Computer Science, October.*
[10] Seitz, C. L. (1985). The cosmic cube. *Commun. ACM* **28**, 22–33.
[11] Snyder, L. (1982). Introduction to the configurable, highly parallel computer. *IEEE Computer* **15**, 47–56.
[12] Bokhari, S. H. and Raza, A. D. (1984). Augmenting computer networks. *ICASE Rep.* 84-33, *NASA Langley Research.*
[13] Despain, A. M. and Patterson, D. A. (1978). X-tree: a tree structure multiprocessor computer architecture. In *Proc. 5th Symp. on Computer Architecture, April.*
[14] Kaisler, S. (ed). (1984). *September Newsletter of Computer Architecture Technical Committee, IEEE Comp. Soc.*
[15] Hoare, C. A. R. (1978). Communicating sequential processes. *Commun. ACM* **21**, 666–677.
[16] Reeves, A. P. (1984). Parallel PASCAL: an extended PASCAL for parallel computers. *J. Parallel and Distributed Comp.* **1**, 64–80.
[17] Snyder, L. (1984) Parallel programming and the Poker programming environment. *Computer* **17**, 27–36.
[18] Li, H., Wang, C. and Lavin, M. A. (1985). Structured process: a new language attribute for better interaction of parallel architecture and algorithm. In *Proc. 1985 Int. Conf. on Parallel Processing, St Charles, Illinois, August.*
[19] Uhr, L. (1984). *Algorithm-Structured Computer Arrays and Networks.* Academic Press, New York.
[20] Kung, H. T. and Leiserson, C. E. (1980). Algorithms for VLSI processor arrays. In *Introduction to VLSI Systems* (ed. C. Mead and L. Conway), pp. 271–292. Addison-Wesley, Reading, MA.

Chapter Six

Considerations on Parallel Solutions for Conventional Image Algorithms

P. Gemmar

1 INTRODUCTION

The majority of algorithms and procedures for automatic evaluation of images have been developed and realized on sequential computers with conventional programming languages. For this and other reasons, image algorithms and operations are typically characterized by heterogeneous processing structures with serial flow of data. In this context image algorithms describe complete computational tasks which evaluate images to get a desired final result. They often consist of procedures combining local and regional image evaluation with object-dependent control of processing. The number of pixels per image to be processed and the rapidly growing number of images acquired, however, often do not allow practical implementations of approved image algorithms on sequential computers owing to their limited processing capabilities. Even simulation or verification of new algorithms on conventional computers can be so time consuming that new ideas sometimes are being dropped before realization.

Basic data structures, data dependencies and operation structures in image processing often possess qualities that can directly be exploited for parallel computing structures to speed up processing [1]. In image-processing, parallelism is available at different levels (e.g. bit level, neighbourhood level, image level, operation level [2]), and can be used in different ways for parallel processing. Therefore various parallel architectures have been proposed, and some parallel-processing systems have been built to speed up image processing. Different processing structures (e.g.

pipeline, array, bus-oriented) with different control structures (e.g. SIMD, MIMD) [2,3] have been developed, with the aim of achieving the highest performance at optimum efficiency for a selected class or classes of image operations with a specific computing structure. However, many parallel-processing systems have severe restrictions on implementation and performance when used for other types of image-computational task that their architecture does not optimally match. Unfortunately, image algorithms often use sets of image operations that require totally different computing structures to automatically evaluate or analyse images. For this reason, a great deal of practical implementation of image algorithms (e.g. for medical image evaluation) has been done with specialized systems adapted and dedicated to the given computational problem.

As will be described in this chapter, image algorithms that have been realized conventionally can hardly be transposed into parallel computing structures. It often takes great expense to do so, and they then show relatively low performance gain with most parallel-processing systems. Although conventional algorithms partly allow parallel processing at operation or object level, complex multiprocessor architectures with both flexible control and data structures are required for implementation [4]. Obviously, parallel processing at data level is one of the most promising approaches in the field of parallel image processing. New problem solutions with parallel algorithms have to be developed using a generalized parallel computing structure which can be implemented on various regular hardware and software architectures. A concept will be described that consists of a flexible structure for iconic image operations and a multilevel data structure for parallel image processing. First results of practical investigations are given for extraction of roads in aerial images.

2 INVESTIGATION OF CONVENTIONAL IMAGE ALGORITHMS

2.1 Operations and computational structures

Image algorithms as defined earlier can consist of primitive image operations, but can also themselves include other image algorithms. Image operations can be classified by their local evaluation and control of processing. We can distinguish between parallel and sequential, homogeneous and non-homogeneous types of local image operations [5]. Homogeneous image operations are described by a uniform function and by parameters with a position-independent selection mask for every pixel. Parallel operations can be performed independently at every input pixel and

Parallel solutions for conventional image algorithms 85

for every output pixel at the same time. Sequential operations are processed in time-sequential order, which is given by a chosen sequence. Homogeneous and non-homogeneous parallel image operations are candidates for effective parallel processing at data level. These types of operations can be found at all levels of image processing, where we can have low-level, intermediate-level or high-level processing. Although parallel operations are mainly expected at low level, typical application algorithms will also use sequential operations in low-level and parallel operations in high-level processing.

In order to make reasonable suggestions for parallelization of image algorithms, one needs precise information about all processing requirements of the complete image-computational task. For this reason, the complete algorithm must be investigated, including all its operations, data and control structures as well as flow of information. By two representative examples it will be shown which possibilities are available to speed up conventional and complex image algorithms by the use of parallel processing. Moreover, processing and structural requirements for potential parallel-processing systems can be extracted without going here into details about data formats, data size, data storage and access or data input/output.

The first example describes an algorithm for separation of objects in multispectral images [6]. It is used for segmentation and description of aerial images. Figure 1 shows the system which is based on a controlled region-growing algorithm. Starting from origin cells of an object to be separated, it accumulates neighbour cells until the object is completely separated. Most of the processing steps in Fig. 1 are independent units and can be isolated for investigation.

(a) *Data reduction.* Principal components transformation is used to concentrate information of statistically independent channels into fewer channels. Determination of covariance and transformation matrices is required for the transformation.

(b) *Extraction of structure elements.* Region growing is controlled by structure information contained in the image. Gradient and contour information in different channels are evaluated separately or jointly and are used to generate a binary edge image which approximates to the region boundaries.

(c) *Distance transformation.* A distance image is generated from the edge image, pixels being thus labelled with their minimum distance to an edge bounding the region containing them.

(d) *An origin cell of one region* consists of all pixels within the region that have a minimum distance (e.g. greater than a threshold) to the bounding edge of that region.

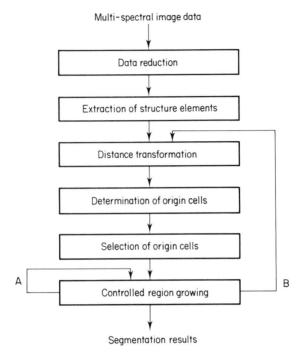

Fig. 1. *System for multichannel separation of objects.*

(e) *Selection of origin cells.* To enhance reliability, only selected cells are considered for the region-growing process. Preclassification of cells is achieved by extraction of geometrical and statistical features of the cells.

(f) *Controlled region growing.* This is the object-separating process as such. Features of neighbouring cells are extracted and compared with those of the origin cell undergoing growth. Neighbouring cells are pixels of a local neighbourhood (e.g. 3×3) along the expanding shell of an origin cell. They should not be confused with cells from (d). Neighbourhood cells will be accumulated, rejected or not accumulated, respectively, until no further cells can be accumulated along the contour of the origin cell. It is important to note that features are recalculated after accumulation of a new shell to the origin cell. In the realized system the region growing process is applied sequentially to cells of one generation (loop A in Fig. 1). New cells will be determined when one generation is finished (loop B).

Parallel solutions for conventional image algorithms

As mentioned before, many of the processing steps (operations) within this system can be inspected separately and independently. This is not allowed for feature extraction and accumulation of cells, where interactions of operations within the system must be considered. The types of operation and their data requirements are listed in Table 1. As we can see, the system for automatic object segmentation in aerial images is mainly built up by a combination of homogeneous parallel and non-homogeneous sequential operations. The region-growing process is the most time-consuming part of the system. This results from the accumulation process as such, with its expense for data management and access, because new origin cells and their feature data are repeatedly processed and generated.

The second example of conventional image algorithms is the extraction of line-shaped objects from aerial images [7]. The computing structure of this system is quite different from that previously described. Its leading strategy of evaluation is strictly sequential and consequently adapted to the shape of the searched objects (Fig. 2). The system integrates two different methods (local and regional) which complement each other perfectly. The basic idea of the extraction method is the analysis of one or more locally limited grey-level functions. Processing proceeds on an object-guided basis from already-known line segments (e.g. roads and rivers). Most-probable locations, orientation and properties of possible neighbour elements are predicted step by step. The system starts with a procedure for the recognition of starting points for the automatic extraction of linear objects like, for example, roads. Beginning with a starting point and knowledge about this object part (e.g. location, orientation, width, contrast) the local method follows the line object on a step-by-step basis. When larger gaps of an object or heavily distorted objects cannot be followed by the local method the system switches over to the regional method. After bridging more complicated parts of an object the system switches back again to the simpler and faster local method.

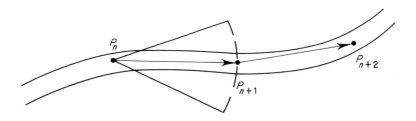

Fig. 2. Sequential road-extraction method.

Table 1
Sample of types of image operation.

Operation	Operation type	Type	Data access	Size
Data reduction	Homogeneous–parallel	Floating-point	Multidimensional sequential	1–3 MB
Gradient extraction	Homogeneous–parallel	Integer	Multidimensional sequential	0.5–1 MB
Contour following	Non-homogeneous–sequential	Integer	Sequential/random	0.25–1 MB
Distance transformation	Homogeneous–sequential	Integer	Sequential	0.25 MB
Determination of origin cells	Homogeneous–parallel	Integer	Random	0.25 MB
Region growing	Non-homogeneous–sequential	Floating-point	Random	1 MB
Recognition of starting points	Non-homogeneous–sequential	Integer	Sequential/random	up to 64 MB
Local method	Sequential	Integer	Random	up to 64 MB
Regional method	Sequential/parallel	Integer	Random	up to 64 MB

Parallel solutions for conventional image algorithms 89

(a) *Recognition of starting points.* Several sample lines and columns of the image are analysed by a profile-analysis operator. This operator performs a special analysis of a one-dimensional grey-level function. For each point of the function it produces a profile characteristic value. Then the sample lines are partitioned and only those parts are accepted as candidates for starting points where features comply with predetermined object features.

(b) *Local method.* With the knowledge about already extracted object parts a possible continuation of the object is predicted and tested via grey-level diagram analysis in a one-dimensional sample cut across the object. The increment of the line object is accepted, if sample features comply with actual object features. The object features will be adapted after acceptance of a new object point.

(c) *Regional method.* This method works in a predetermined limited area of the image with usually rectangular shape. This area is expected to completely contain a part of the line object. The area is partitioned into sample lines which possibly contain cross-sections of the object. These sample lines again are subjected to a grey-level analysis to discover candidates for object points. These candidates then are examined in order to distinguish between isolated points and points that contribute to form a line.

Regarding the image operations and algorithms discussed so far, the association of types and computational structures to certain operations is sometimes difficult. Let us take for example a contour-following algorithm and the local method for extraction of linear objects. Both algorithms can be described as non-homogeneous sequential algorithms. They both evaluate data in a relatively small local neighbourhood and then decide which data are to be processed next. But the difference lies in the operations carried out. In the case of the contour-following algorithm one can speak of an application of an image operator. In contrast, the road extraction algorithm is a complex task with complicated control mechanisms and which needs several thousands of instruction lines to be coded in a high-level language. The implications are apparent for hardware realizations of both systems.

Another hidden problem is related to the partition of image algorithms into individual processing steps or operations. Image algorithms cannot always be partitioned into single operations without gaps. On the contrary, there often remains a residue of control business which cannot be classified. This portion of the job varies with each algorithm and should not be neglected. It can play an important part when parallel processing of an object or operation level, for example, is considered to speed up processing.

Reduction of speed-up rates by significant factors must be taken into account for many demanding image algorithms which were not designed for parallel processing from the very beginning.

2.2 Effectiveness of some methods of parallel processing

In this section different methods will be discussed for speeding up image operations, and their effects will be investigated when applied to previously described image algorithms. First we shall look at some speeding-up actions that have little relation to specific computing architectures. Then the possibilities and influence of parallel processing at different levels of image operations or algorithms respectively will be described.

2.2.1 Instruction execution and data access

The actions to be described in the following are general methods which can be used to directly speed up image processing. Therefore they can be directly exploited without too great expenditure by modern general-purpose computers, as well as by specialized processing systems.

(i) *Microcode.* It is common practice for instruction sets to be realized and executed under microprogram control. Also in this way, particular routines can be microprogrammed and provided as fast machine instructions. This can be worthwhile for many image operations that are frequently and repeatedly used. The achievable gain of processing speed varies and depends on the routine being microprogrammed. Speed-up factors of up to 10 are theoretically achievable; in practice, factors between two and five have been measured on a VAX 11/780.

(ii) *Address calculation.* Usually, images are scanned, digitized, and then stored as square matrices of point values (pixels). The linear size of the matrix is typically measured in powers of two. If a matrix of size N is stored in memory at base address S, access to pixels $B(i, j)$ is given by their memory address:
$ADR\ (B(i, j)) = S + i{\star}N + j \quad (i,j = 0,1, \ldots, n-1).$

In many simple image operations, address calculations take the major proportion of processing time, especially when they are performed on non-array-related architectures or systems that do not have any provision for effective pixel access. Independently of this, address calculation is often required in sequential image operations and algorithms that need random access to pixels during processing. However, address calculation can be drastically simplified if images are stored in memory starting with base addresses in terms of the image size. By this method, pixel addresses can simply be composed of base, column and row addresses without time-consuming multiplication/add operations. In practice this method of storage allocation and pixel addressing can speed up image operations by factors of up to five, with only small additional costs in hardware and software, and there are also no structural conditions for the image operations to be performed.

(iii) *Memory data access.* Most of the known image operations process locally limited two- or three-dimensional neighbourhood image data in order to generate new image or feature data. In the case of two-dimensional neighbourhood image data, a submatrix or window U of size $n \times n$ is normally taken to produce a local result. Certainly, it is not surprising that there are many different system architectures that are optimized to perform local neighbourhood operations. In general, they try to achieve direct access to neighbourhood pixels by a variety of techniques [8]. However, they also show a common drawback in their restricted ability to perform sequential operations. In contrast with parallel operations, sequential operations can determine the next position of a local window U only after processing of the previous one is finished. Nevertheless, the region can be predefined where the image data to be processed next will be found. Given a sequential algorithm with step width d to the next position of window U, new data will be found in submatrix U'. One obtains submatrix U' on enlargement of submatrix U by border sectors of width d. During processing, data of the window U' are stored in a fast cache memory. Window U' is continuously updated by reloading the border sections into which submatrix U has moved during processing. Establishing a cache memory for window U' will therefore provide a simple means of at least doubling the processing speed for sequential and parallel operations.

The methods described so far can be utilized to speed up every type of image operation/algorithm on various types of computational architectures. Together with modern high-speed hardware technologies (e.g. GaAs), flexible and fast image-processing systems can thus be built at reasonable costs for a great variety of applications.

2.2.2 Parallel processing at different processing levels

As mentioned in Section 1, there are many good reasons to speed up image processing. Enhancement of today's available processing power by some orders of magnitude is demanded for many practical applications. One way to continue could be the development of new and clever image algorithms that save computing power, but in most cases advanced algorithms for image processing show higher complexity, with increased computing requirements. In consequence it seems that the only possible way to speed up image processing is by the use of highly parallel computing structures. As was described earlier, parallel processing can be performed at different levels. Now the question arises, to what extent can conventional algorithms be processed on parallel systems? Parallel-processing systems with m processing elements or m processing stages can theoretically speed up processing by a factor of m. What rate can be expected with practical implementations? In the following the most important levels of parallel processing for image processing will be briefly described.

(i) *Data level*. Parallel image operations allow the processing of all pixels independently and at the same time. Partitioning of images into subimages is a special but more demanding type of data parallelism. The highest degree of parallel processing is exploited by array architectures. Non-homogeneous parallel operations can be realized with specialized data and control structures.

(ii) *Operation level*. An image operation (and others) can be partitioned into a set of primitive operations or instructions which can be executed independently and in a given order. This set of operations can be arranged in appropriate parallel-processing structures like a pipeline or a cascade. Different operations use different structures, which need a flexible and a structural programmable multiprocessor system for optimal implementation [9]. Best results will be achieved with image operations that are constant for the whole image (data parallelism), and which can be partitioned into primitives of equal execution load.

(iii) *Function level*. Many image algorithms contain procedures that can be executed independently. The cooperation of the procedures can show static or dynamic parallelism. The former type of parallelism is defined at the programming stage. The latter emerges during task execution (e.g. extraction of several roads starting from crossroads). Construction of a pipeline with complete procedures at every stage is

Parallel solutions for conventional image algorithms

called "macropipelining". This method can be used efficiently for processing image sequences without data feedback.

In the previous two sections six different methods have been described for speeding up image processing. The first three methods can be used with arbitrary processing systems. The last three methods require multi-processor systems to be implemented. Therefore speed-up rates depend on the number of processing elements (processors) available. On the other hand, conventional algorithms provide different degrees of parallelism which can be converted into parallel processing. The performance of different parallel processing methods has been measured for the two conventional image algorithms described previously. Tables 2 and 3 show the different results obtained for the different methods. Parallel processing at data level is only available within intermediate processing steps of the algorithms. It can be seen that parallel processing at the operation level does not show good results with complex image algorithms. Higher speed-up factors SF with m parallel units are only achievable if the time proportion p (in percent) of an operation reaches nearly the total time for the task:

$$SF = 100/(p/m + 100 - p).$$

Parallel processing at the function level increases p and can also parallelize particular control functions and data management operations. Naturally, this type of processing assumes appropriate system capabilities for data processing, data storage and process control.

3 SOME CONSIDERATIONS ON PARALLEL SOLUTIONS FOR HEURISTIC IMAGE OPERATIONS

The investigation of conventional image algorithms in the previous sections showed that this important kind of heuristically developed image process imposes critical conditions for parallel processing. Their defined computing structure does not make feasible an effective usage of the technically most simple parallel processing at the data or operation levels. A method for converting conventional algorithms systematically into advanced parallel-processing structures is not available. There are several reasons why given algorithms in many cases are rather impracticable for parallel processing, for example, hidden parallelism, mixed organization of sequential and parallel operations, random data access with distributed storage of data, and extensive intermediate process control. Only by the use of so-called "general-purpose image multiprocessors" will it be possible to exploit parallelism of image algorithms in an efficient way, as described in the

Table 2
Speed-up rates for the image segmentation task.

Parallel processing-level	Parallel proportion of task	Speed-up rate with m processors	
		m = 16	m = 128
Function level: growing of multiple cells	80%	4	4.85
Operation level: parallel accumulation of one cell	42–64%	1.65–2.5	1.7–2.74

Table 3
Speed-up rates for the road-extraction task.

Parallel processing-level	Parallel proportion of task	Speed-up rate with m processing units	
		m = 16	m = 128
Function level: parallel processing of subimages	Up to 100%	Up to 16	Up to 128
extraction of multiple roads	Up to 100%	Up to 16	Up to 128
Operation level: parallelization of profile analysis	20%	1.23	1.25

previous section. But until now, such systems have not been available, and it may be asked whether by existing and specialized parallel image-processing systems the cumbersome mental gap can be closed between the sequential way of thinking and parallel solutions for image processing.

3.1 A constituent concept for parallel iconic operations

Since there is hardly any way to implement parallel image-processing tasks other than by developing new parallel algorithms, a general consideration of parallel image operations should be made. By the principal formulation of parallel operations their implementation will be possible on nearly any parallel architecture. At least, transformation of described operations is possible for different implementations on different architectures. Combina-

Parallel solutions for conventional image algorithms 95

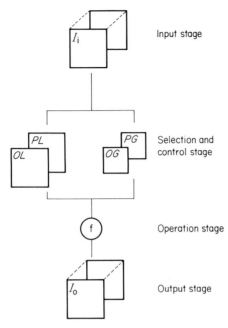

Fig. 3. Structure of parallel iconic operations.

tion and composition of operations can be provided by a set of constituent primitives. A structural concept for parallel iconic operations is given in Fig. 3.

Iconic operations as graphically described in Fig. 3, for example, can be executed locally independently and with local or global process control. For this reason they have an input stage with image and mask data, and a selection stage with appropriate data structures for local or global selection and control of processing. As mentioned before, different processing architectures can be provided for the processing stage, and the output stage can comprise multiple data sets just like the input stage. A parallel iconic image operation PI is therefore described by its basic function f, its input data (list) (I_i), and its optional declarations for local (OL, PL) or global (OG, PG) modification or selection of the operation and parameters, respectively. The output data (list) (I_o) is given by $I_o = (PI)(I_i)$, with $PI = \{(f,I_i) \mid (OG, PG) \wedge (OL,PL)\}$.

3.2 A parallel approach to a sequential image algorithm

The extraction of roads was chosen for practical implementation of a parallel image algorithm. Although the previously described successfully working sequential algorithm exists, the primarily non-parallel image-processing problem required a new system design for parallel processing. The previously described structure of iconic operations has been used for development of parallel operations. A multilayer data structure has been generated and used. The flexible data structure contains different layers of information matrices with different size. By this means the operations described above can be used, for example, to control processing locally, to provide a flexible selection and condition logic, and to establish combinatorial feature-extraction methods. A simulation system is being built for the development and verification of parallel image algorithms. The current system comprises several parallel operations for the extraction of line objects (roads) from aerial images. The operations can work independently, iteratively, or together in a given order. Some of the operations and their usage for a practical example will be described briefly in the following.

LF This is a local function on selected neighbourhood elements and with a global declaration of parameters. *OG* is used to select neighbour elements, *PG* to provide global weight factors or parameters (e.g. $PI = \{f(I_i, PG(p,q)) \mid (p,q) \in OG\}$).

ED This is a locally adaptive operator for extraction of edges from linear objects. It analyses a small (e.g. 3×3 or 5×5) binary-valued window in order to detect an edge point and its corresponding direction. Binary windows are obtained by comparison of grey-valued pixels with a local threshold already computed by operation *LF*. *ED* uses *PL* (same size as I_i) to provide the local thresholds, *OG* to select the interesting elements in the binary window, and *PG* to describe allowed binary sets of windows for the determination of edge points.

DP This operation was developed to extract road points and their directions. Given a grey-valued image and its corresponding image with edge points extracted by *ED*, operation *DP* will determine directed road points where there are two antiparallel edge points in an allowed distance, with appropriate grey values on each side. *DP* uses *OG* to define local search areas for each edge direction, and *PG* to establish allowed directions of possible partner points.

LO The length of directed line objects is evaluated by *LO*. Depending on its actual direction, two locally selected neighbour points of a line point are evaluated to determine its local length. Iteration of *LO* will supply each object point with the total length of the line object. *OG* and *PG* are used to select the neighbour points and their allowed direction for a given object point.

Parallel solutions for conventional image algorithms 97

CO, There are two complementary operations to compress or expand an
BE image with directed object elements into images with lower or higher resolution respectively. The operations consider the local and specific features (e.g. direction and length) of object points during evaluation. For this reason PL will contain (for every object point) local information about the length of the object to which the point belongs. As described before, OG and PG again are used for the selection of neighbour elements and their allowed values.

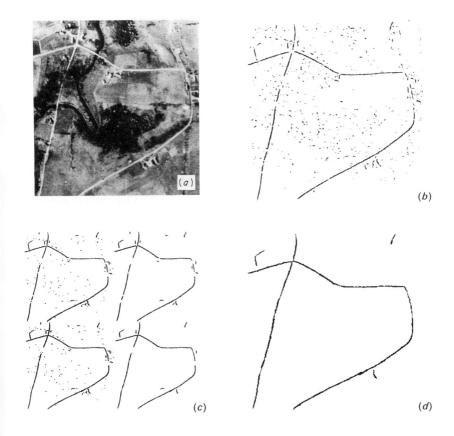

Fig. 4. *Extraction of line objects from aerial images. (a) Aerial image, 512*512 pixels. (b) Extracted direction elements of line objects (equally valued for display purposes). (c) Top left: extracted candidates for road points at lower resolution; top right: selected "good" line points; bottom left: bridging of gaps between line points; bottom right: extracted steady line points. (d) Stable road parts after iteration of steps 3 and 4.*

An example will now be given for the practical application of the parallel algorithms described above. Extraction of steady road points from an aerial image is carried out by a flexible combination of operations in a multilevel data structure. In general, one can distinguish between four processing steps:

(1) detection of edges from line objects (roads);
(2) detection of directed object points (road points);
(3) determination of stable object points;
(4) bridging of object parts with stable points.

In the first processing step a filtered (edge-preserved) image will be processed in sequence by operations LF and ED to produce an image with directed edge points. In the second step operator DP is used to extract possible points of line objects. Figure 4(b) shows a result of these operations applied to Fig 4(a). Operation LO is used in the third step to evaluate the length of existing object fragments (Fig. 4(c)). The actual length of object parts is used for selection of "good" line points and also as a condition within the operations CO and BE. Connection of good object parts with other (shorter) ones is performed in the fourth processing step. At the last two stages images of different resolutions are produced and examined. Operations BE and CO are used to expand and compress images of line objects in a manner that is sensitive to the direction of object points. In order to bridge further gaps between object parts, processing steps three and four can be performed iteratively (Fig. 4(d)).

4 CONCLUSIONS

It seems not to be practical for conventional and approved image algorithms to be executed in a parallel manner by the use of their inherently available parallelism. In most cases they provide partly object-driven and complex computing structures with considerable requirements for process control, which do not make it feasible to use regular hardware and software architectures for suitable acceleration of processing. A general-purpose multiprocessor system for image processing with a lot of powerful processing units, data memory and interconnections is not to be expected in the near future owing to unsolved hardware and mainly software problems.

Parallel-processing structures like pipeline, array or pyramid systems are proposed to make reasonable use of advanced potentialities of modern technology. However, new image algorithms are required that offer data parallelism for parallel processing. Parallel image operations need a generalized operation structure with flexible processing capabilities. Implementation of parallel algorithms should be possible on different system architectures without essential reformulation of the process. A concept for parallel iconic operations with a generalized processing structure was described. Based on this structure, a simulation system for parallel operations was developed. Extraction of roads from aerial images was chosen for first application to a basically non-parallel image algorithm. Parallel operations for extraction of line elements have been discussed and results for extraction of steady road parts have been shown.

REFERENCES

[1] Unger, S. H. (1958). A computer oriented toward spatial problems. *Proc. IRE* **46**, 1744–1750.
[2] Danielsson, P. E. and Levialdi, S. (1981). Computer architectures for pictorial information systems. *IEEE Computer Magazine* (Nov.), pp. 53–67.
[3] Duff, M. J. B. (1982). Special hardware for pattern processing. In *Proc. 6th Int. Joint Conf. on Pattern Recognition, Munich*, pp. 368–379.
[4] Gemmar, P. and Palomino, F. (1981). Untersuchung von Prozessorstrukturen fuer die Bildverarbeitung. *BMFT-Forschungsbericht DV 81–008, Fachinformationszentrum Karlsruhe, Germany*.
[5] Kazmierczak, H. (ed.) (1980). *Erfassung und maschinelle Verarbeitung von Bilddaten*. Springer, Wien, New York.
[6] Schaerf, R. (1981). Untersuchungen zur bildgesteuerten Separierung von Objekten in multispektralen Daten. Dissertation, Universität Karlsruhe, Germany.
[7] Bausch, U., Bohner, M., Groch, W.-D., Kazmierczak, H. and Sties, M. (1981). Automatic extraction of linear features from aerial photographs. *Final Tech. Rep., European Research Office London, England*.
[8] Preston, K., Duff, M. J. B., Levialdi, S., Norgren, P. E. and Toriwaki, J. (1979). Basics of cellular logic with some applications in medical image processing. *Proc. IEEE* **67**, 826–856.
[9] Gemmar, P., Ischen, H. and Luetjen, K. (1981). FLIP: a multiprocessor system for image processing. In *Languages and Architectures for Image Processing* (ed. M. J. B. Duff and S. Levialdi), pp. 245–256. Academic Press, London.

Chapter Seven
Connected Component Labelling in Image Processing with MIMD Architectures

R. Hummel

1 INTRODUCTION

Many low-level vision tasks are ideally suited for parallel computers configured as a large grid of locally connected simple processors executing a single instruction stream. Such SIMD (Single Instruction, Multiple Data) image-processing computers now exist [1–3]. Common tasks such as point operations, thresholding, convolution, texture analysis and median filtering are easily transferred to such an architecture. A range of other tasks, such as morphological operations (shrinking, expanding, smoothing), region growing, and relaxation operations for segmentation are also frequently well suited to mesh arrays. However, many of these operations are really ways of identifying structure within the image, and to make measurements in particular locations within the image, by treating each pixel in a uniform way. That is, on a mesh-connected architecture, these algorithms in fact perform useful work on only a fraction of the image pixels. Finally, for many higher-level vision tasks, such as boundary parameterization, feature extraction, object matching and symbolic constraint propagation, mesh-connected architectures can at best clumsily provide the necessary data-communication paths.

Some of the mesh-connected machines are enhanced with global operations, such as a "sum-OR" or even a global sum of bits, in order to give the machine the ability to perform fast image maxima or histograms. This increases the ability to extract Hough transforms and other image-wide features. The difficulty here is that region-based features and local Hough transforms can only be accomplished by masking out the relevant regions, and operating serially on the individual regions.

Certain algorithms are necessarily nonlocal. For example, connected-component extraction and convex-hull operations on binary images require long-range communication, as pointed out in [4]. Although these can be implemented on mesh-connected machines, typically a large number of processing steps are required. For example, for connected-component labelling, the standard algorithm is equivalent to a breadth-first-search, and works by having pixels within each region iteratively replace a current identifier with the maximum of identifiers within its local region. If each pixel begins with a unique identifier, then eventually each component will be labelled with the number representing the maximum identifier in that component at the start of the process. An asymptotically faster approach is available, which works by a divide-and-conquer approach [5]. Connected-component analysis is done within each of four quadrants of the image, by a recursive procedure, and then the components are combined by an equivalence table analysis, done within the mesh, along the interface borders of the quadrants. This latter algorithm can work in time $O(n)$ on an n by n grid. On a pyramid machine, the worst-case complexity can be further reduced to $O(n^{\frac{1}{2}})$ [6].

Once components are extracted, it is usual in image processing to analyse those components, and form complex representations, in order to perform recognition and scene analysis. Features like moments, Fourier descriptors of the boundary, other curvature measurements, three-dimensional shape analysis, and graph structure of regions generally argue for a standard, powerful processor capable of accessing all the image data. Parallel architecture design would then suggest that if one processor with global access is useful for higher-level vision analysis, then two processors will be more advantageous. In fact, if the standard image has a few dozen or a few hundred regions whose shapes, descriptions and local context require analysis in relative isolation from the remaining portions of the image, we would like to have available several dozen or hundreds of processors, each capable of running an independent instruction stream, and each reconfigurable in the sense that it can be assigned to a variable portion of the image. We are thus led to consider MIMD (Multiple Instruction, Multiple Data) shared-memory architectures for image processing.

Parallelism of the order of a few hundred or thousand processors in MIMD shared memory mode offers a number of attractive features for image processing. Since the architecture can be considered "medium-grained" (granularity is relative; here we are comparing with the number of pixels in a typical image), each processor can be a quite capable multi-MIPS (Million Instructions per Second) microprocessor, with a rather complete

Connected component labelling 103

instruction set. In particular, we see each processor as having registers, a local cache memory, as well as complete access to global memory. This compares with fine-grain, typical SIMD architectures, where the processors have access exclusively to a few hundred bytes of local memory, and are typically bit-serial or 8-bit processors with an extremely limited instruction set. The use of powerful processors eases algorithm development and enhances flexibility in design and coding. We should expect that nearly all standard image processing algorithms can be handled with relatively good efficiency and programming ease on a multiprocessor shared-memory design.

Further, reconfigurability of the assignment of processors to image regions coincides with notions of "focus of attention" and foveal resolution versus periphery processing. Each processor can work on a specific component, communicating with neighbouring components through the relevant processors by means of an "edge adjacency list" data structure for the dynamic graph of image regions.

Finally, since processor power can be targetted to relevant portions of the image, we expect to increase the degree of parallelization. In grid arrays, with fixed assignment of processors to pixels, many of the processors perform no useful work during execution of a cycle, simply because they are statically assigned to locations where work is not needed. Thus many algorithms yield degrees of parallelization in the range of a fraction of a percent. With the reconfigurability of a medium-grain architecture, we expect to increase the usage figure, at least for higher-level image processing tasks, to the tens of percent.

There are now a number of proposed and prototype MIMD shared-memory computers [7,8]. The use of these machines for image processing has received some attention [9,10]. However, the conversion of standard algorithms to MIMD parallel code offers some subtleties. Very high-level vision tasks presumably offer few obstacles, because little coordination between processors is necessary. Sharing of data is usually minimal. However, for lower-level vision tasks, there are difficulties related to synchronization and assignment of processors. In this paper, we study as an example case the graph-theory algorithm of Shiloach and Vishkin for connected component labelling [11]. This algorithm, like most intermediate-level vision tasks, is presented as an SIMD parallel algorithm. We study how such an algorithm can be mapped to a realistic MIMD machine. Our model for the MIMD machine is the NYU Ultracomputer [8]. Our choice of the connected component algorithm is based on the observations that component labelling is necessarily a non-local operation and that the $O(\log N)$ algorithm is non-obvious but easily described and coded.

There are two main concerns that arise in converting SIMD parallel algorithms to MIMD medium-grain architectures. First, it is usually possible to concentrate processors at locations where useful work will be accomplished. We allow available processors to perform a step in the algorithm, and simultaneously create a list of locations to be considered in subsequent steps. In much the same way that data-flow techniques analyse the flow of data through processing steps in an algorithm, MIMD architectures in image processing demand a "task flow" analysis. The data can be regarded as fixed, like a blackboard, but the tasks that must be assigned flow in a way which depends upon the initial data. Consider, as a trivial example, an erosion operation on a binary image. Erosion, for our purposes, can be defined as setting all boundary pixels with value "1" to "0", thereby eroding the border or regions. The SIMD approach to this operation is to replace each pixel's value with the minimum in the local neighbourhood. However, in the SIMD approach, we note that "useful" work is accomplished only on the border of regions. The MIMD approach would therefore be to create a list of points on the borders of regions, and to assign processors to the tasks of resetting those pixels to the value "0". At the same time, processors can determine a new list of boundary points, where the next iteration of an erosion operation will create new pixel values. The new list might contain redundancies, but it is nonetheless a considerably smaller list than the set of all pixels. In this algorithm, we see that the tasks are assigned to border pixels, and that the border flows in a predictable way, such that a new list of border points can be constructed from an iteration applied to previous border pixels.

A second concern is one of synchronization. The SIMD algorithms assume that many tasks are accomplished concurrently, and normally assume that concurrent reads are allowed and yield identical information. Frequently, concurrent writes are also permitted, and it is generally assumed that exactly one write to a specific location at a given time is successful. When we convert to an MIMD environment, in image-processing applications we are normally in a processor-poor situation. Thus tasks that were meant to be done concurrently will be at least partly serialized. Standard issues of data concurrency arise: if early serialized subtasks change the values of data, then later tasks may access data that is different from what would be expected in a synchronous application of all tasks. These problems can be handled by one of three methods: (i) read from a fixed copy of the data, and write a new copy with appropriate changes; (ii) read from the data, writing changes as a list of tasks to be performed later, after all of the serialized analysis tasks are finished; or (iii) show that asynchronous operation of the steps, which is what partial serialization with early writes to the data constitutes, is a satisfactory approach for the algorithm in question. For the erosion example, the third alternative is

Connected component labelling

probably not appropriate, since changing a border pixel from a "1" to a "0" changes what might be considered a border pixel, so that if the list of tasks contains border candidates that are not border points, a subsequent task might change a pixel which otherwise would be considered an interior point. Of course, if the list of points is always known to be a (possibly redundant) list of border points, then the tasks amount simply to setting values to zero, and determining new border points, and asynchronous operation will work. Otherwise, the second approach, of creating a list of border points to be converted to "0"s and to enqueue new candidate boundary points on a new list, will be necessary.

2 A PARALLEL ALGORITHM FOR CONNECTED COMPONENTS OF A GRAPH

Shiloach and Vishkin [11] give an algorithm to find connected components of an undirected graph using a computational model which allows simultaneous reads and writes by multiple processors to a common memory. In the case of simultaneous writes, the model specifies that one of the writes succeeds, but does not specify in advance which processor will be successful.

Each vertex in the graph is assigned a *parent* pointer. The algorithm uses these pointers to construct a forest of root-directed trees, also called a *pointer graph*; each tree is composed of vertices known to reside in the same connected component. Initially each vertex is its own parent, i.e. each vertex comprises a rooted tree with only one node. At termination, the parent pointers of all vertices in the same connected component point to the same root vertex.

In the Shiloach/Vishkin parallel algorithm, each vertex is assigned a processor. Each edge is assigned two processors; in effect, a processor in each direction of edge traversal. Two operations are performed in the algorithm. The *shortcut* operation redirects a vertex's pointer to the parent of its parent, reducing the height of the trees. Shortcutting is mediated by the vertex processors. The *hook* operation directs the pointer of the root of a tree (self-directed before the operation) to point to some node in another tree. Hooking is mediated by the edge processors. We let V be the set of vertices and E the set of directed edges, one for each edge processor.

At intermediate stages of the algorithm, the trees in the pointer graph can be classified as *live*, *stagnant* and *dead*. A live tree has been shortcut or hooked in the current iteration. A stagnant tree has not been shortcut or hooked in the current iteration. A dead tree has not been shortcut or hooked in the previous iteration.

In the algorithm that follows, each vertex is assigned an *age* value, which is used to determine when trees are stagnant or dead. Further, we assume that each vertex v is denoted by a distinct integer. Thus if v is the integer value for a vertex, parent[v] is the integer value of v's parent vertex, and age[v] is the integer value of the last iteration in which v was shortcut or hooked.

```
                /* initialize */
0)      (∀v)[v ∈ V] cobegin
            parent[v]←v;
            age[v]←0;
        coend
        I←0;
        While (∃v)[age[v] = I] begin
            I++;
            /* Shortcutting */
1)          (∀v)[v ∈ V]cobegin
                old-parent←parent[v];
                parent[v]←parent[parent[v]];
                if (old-parent ≠ parent[v]) age [parent[v]]←I;
            coend
            /* Ordered Hooking */
2)          (∀ (x,y))[(x,y) ∈ E] cobegin
                If parent[x] = parent[parent[x]] /* x points to a root */
                and parent[x] < parent[y] then begin
                    parent[parent[x]]←parent[y];
                    age[parent[y]]←I;
                end
            coend
            /* Stagnant Hooking */
3)          (∀ (x,y))[(x,y) ∈ E] cobegin
                if parent[x] = parent[parent[x]] /* x points to a root */
                and age[parent[x]] < I /* root is stagnant */
                and parent[x] ≠ parent[y] then begin
                    parent[parent[x]]←parent[y];
                    age[parent[y]]←I;
                end
            coend
            /* Shortcutting */
4)          (∀v)[ ∈ V] cobegin
                old-parent←parent[v];
                parent[v]←parent[parent[v]];
                if (old-parent ≠ parent[v]) age[parent[v]]←I;
            coend
        end
```

Connected component labelling

3 TIME COMPLEXITY OF THE ALGORITHM

The correctness of the algorithm is easily established. We shall concentrate on the time bound. The proof sketched below differs only slightly from that of [11]. In [11] a bound on the cardinality of the trees is related to the iteration counter I. Here, we show instead a bound on the *height* of the live trees in the forest formed by the parent pointers. We define height as the maximum number of pointers that must be traversed in a simple path from any vertex in the tree to the root, with the additional proviso that tree height is always at least one. In particular, a tree consisting of only a single node (the root node with a self-loop) is considered to have height one, as is a tree where all vertices point to the root.

We first quantify the effect of shortcutting with the following Lemma.

Lemma 1

Shortcutting a tree of height $h > 1$ reduces its height by at least a factor of $3/2$.

Proof For any tree of height $h > 1$ shortcutting yields a height of $\lceil h/2 \rceil$. But for any $h \geq 2$ we have

$$\frac{h}{\lceil h/2 \rceil} \geq 3/2.$$

□

We shall also need a rather technical lemma.

Lemma 2

If $u >$ parent$[u]$ after any step in the algorithm then either u is a leaf node or all of u's predecessors in the pointer graph are leaf nodes.

Proof The proof is by induction on the steps in the algorithm. Initially, since parent$[u] = u$ for all vertices, the lemma holds true vacuously. If the lemma holds true at the start of a shortcutting step (Steps 1 or 4), then the shortcutting operation of pointing u to w where there were links from u to v and v to w can create a new link with $u > w$. However, in this case, either $u > v$ or $v > w$ before the shortcutting. In the former case, either u is a leaf or all its predecessors are leaves. In the later case, v's predecessors, including u, must be leaves. In either case, after the shortcutting, u must be a leaf.

Ordered hooking (Step 2) can never create a link from a larger to smaller vertex.

Stagnant node hooking (Step 3) can link a node u to a node v with $u > v$.

However, in this case, u must be a root of a stagnant tree. A stagnant tree cannot have a height greater than 1, or else it would have been shortcut in Step 1. Thus at the start of Step 3, u is either a leaf, or all of its predecessors are leaves. During Step 3 no other stagnant root can hook onto u, since otherwise in Step 2 one of the two roots would have been eligible to hook onto the other. Thus in Step 2 either u would have succeeded in hooking onto something, in which case it is not a root, or something succeeds in hooking onto u, in which case it is not stagnant. Further, during Step 3, any of u's predecessors will remain leaves, since nothing can hook onto a leaf. Thus at the end of Step 3 the lemma holds for node u. This establishes the induction for steps 1, 2, 3 and 4. □

Lemma 3

If n trees of height h_i, $i = 1, \ldots, n$, are hooked together, the result is a tree of height no more than Σh_i.

$$\sum_{i=1}^{n} h_i.$$

Proof The first part of this lemma is to show that the result is a tree; that is, we must show that no cycles are created. Suppose that a cycle is created after some step. Then somewhere along the cycle there must be a pair of consecutive nodes u and v such that $u > v$ and u points to v. According to the previous lemma, either u is a leaf, or all of u's predecessors are leaves. Either way, u cannot be part of a cycle.

Consider hooking T_1 into T_2, assuming both trees have more than one node. Let their heights be h_1 and h_2 respectively. After hooking, the maximum height h of the new tree is no more than the sum of the heights of the old trees, plus one for the new link, less one since we could not hook to a leaf in T_2; thus

$$h \leq h_1 + 1 + h_2 - 1 = h_1 + h_2.$$

If either or both trees has one node, that node can contribute at most 1 to the height of the resulting tree, so the result follows. The result easily generalizes when n trees are hooked together. □

Definition

A root u is *eligible* if there exists an edge $(x,y) \in E$ and a node $v \neq u$ with parent$[x] = u$ and parent$[y] = v$, such that either $v > u$ or age$[u] < I$.

Connected component labelling

Lemma 4

Eligible roots are always hooked.

Proof Suppose u is an eligible root, and (x,y) is the edge with parent$[x] = u$ and parent$[y] = v$. If $v > u$ then in Step 2 the edge processor (x,y) will attempt to hook u onto v. Either (x,y) succeeds or some other processor succeeds; in either case, u will hook onto some node. If age$[u] < I$ and u does not hook in Step 2 then u is a stagnant root by Step 3, so the processor (x,y) will attempt to hook u to v. Once again, some such processor succeeds, and so u hooks. □

Lemma 5

Dead trees stay dead.

Proof Recall that a dead tree is one that has not been shortcut or hooked in the previous iteration. Clearly the height of a dead tree must be 1, or it would have been shortcut. Also, its root must not be eligible, or it would have been hooked (Lemma 3). Thus it can only be brought back to life if another tree is hooked into it, which implies the existence of some edge (x,y), where x is in the dead tree and y is not. But then the root would be eligible, since u must point the root, which is stagnant by Step 3 of the iteration in which the tree dies, and we have a contradiction. Thus if a tree is dead, it cannot be hooked into or shortcut, so it stays dead. □

We can now present the key

Time bound

The sum of the heights H_i of all live trees at the end of iteration i is bounded by

$$H_i \le \left(\frac{2}{3}\right)^i N,$$

where N is the number of vertices in the graph.

Proof By induction on i. Initially $(i=0)$, each vertex forms its own tree, of height 1, so $H_0 = N$. The induction hypothesis states that after $i-1$ iterations the sum of the heights H_{i-1} of the live trees is bounded by

$$H_i - 1 \le \left(\frac{2}{3}\right)^{i-1} N.$$

We show that a tree in the forest of height h at the end of iteration $i-1$ contributes no more than $\frac{2}{3} h$ to H_i. If $h = 1$ and the root is not eligible, the tree dies, and it contributes nothing to H_i. If $h > 1$ and the tree is neither eligible nor hooked into by some other tree, it is shortcut in Step 4, and its contribution to H_i is at most $\frac{2}{3}h$. If the root hooks or is hooked, then the height of the new tree to which it belongs is no more than the sum of the old heights of the individual trees which have hooked together. However, since the new tree has height greater than 1, and Step 4 will shortcut the tree, its height will be at most $\frac{2}{3}$ the height of the sum at the end of the iteration. Thus each participating tree contributes no more than $\frac{2}{3}$ its height in the previous iteration to H_i. □

As a direct consequence of the bound above, we see that after more than $\log_{3/2} N$ iterations, we must have $H_i = 0$, which is to say that all trees are dead, and the algorithm terminates.

Note that the Step 1 shortcutting is necessary to define eligibility of a node (a stagnant node). Since Step 1 may be applied to trees of height 1 which do not reduce in height and yet do not die because they can hook, we cannot claim a 3/2 factor total height reduction on account of Step 1.

4 NECESSITY OF THE STEPS

As noted in [11], Step 4 is unnecessary, and omitting it will at worst double the number of iterations required for termination. Its inclusion simplifies the proof, and less than doubles the amount of work required per iteration while halving the number of iterations that will be needed in the worst case. Whether the trade-off is reasonable in the average case is best judged by empirical studies.

Step 2 is needed to prevent cycles from forming. If Step 3 is omitted then the algorithm will still work well—however, the $O(\log N)$ time bound is no longer valid. Dropping Step 3 is appealing, however, since it is the only step that introduces pointers to smaller vertices. Without Step 3, the final labelling will be components with vertices pointing to the maximum vertex within each component. If the vertices are numbered according to some labelled value, this provides an algorithm for simultaneously labelling connected components and finding maximum values within components. If

Connected component labelling

the $O(\log N)$ connected components algorithm is performed first, and then maxima found within each component, the result for an Ultracomputer is an $O(C \log N)$ method, where there are C components. If the Ultracomputer permits parallel maxima, in the same way that it permits parallel adds, then the method is $O(C + \log N)$. Thus if omitting Step 3 retains the $O(\log N)$ performance of the algorithm, we have a better component maximum method; although the advantage to the case where parallel maxima are allowed is very slight.

Unfortunately, as is shown in [11], the algorithm without Step 3 can result in $O(N)$ time performance. The situation arises with a *star* graph. We might conjecture that for subgraphs of images, which are planar graphs with bounded in-degree, such examples will not arise. However, Fig. 1 shows a *comb* graph, which when numbered as shown, and under somewhat perverse conditions, will require $O(N)$ iterations. The first hook operation can yield the pointer graph shown in Fig. 2. The only eligible root is $n + 1$, which in the next iteration might hook to $n + 2$. Continuing in this way, a possible sequence of $n/2$ stagnant node hooks will be needed.

Fig. 1. A comb graph.

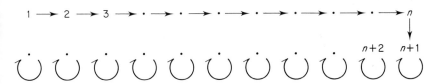

Fig. 2. *The pointer graph after the first min-neighbour hook operation. If stagnant node hooking is omitted, then after a couple of short-cut operations, the only eligible root is "$n + 1$", which might then hook to "$n + 2$". This process can continue, yielding $n/2$ iterations before termination.*

If the graph is ordered in raster-scan fashion, instead of with malicious intent as was done in the example above, it is still possible to construct an example which needs linear time. In this case, however, the comb has to be laid out in one dimension across the image, and so an n by n image is limited to $O(n)$ complexity.

The perverse nature of these counterexamples leads us to conjecture that an expected time of $O(\log N)$ would be obtained by the simplified algorithm omitting Step 3 if all hooks are equally likely.

5 TASK FLOW OF THE ALGORITHM

The proof of the algorithm allows us to make some immediate observations about when processors will be needed for useful work.

Observation 1

All vertices in a dead tree and all edges adjoining vertices in a dead tree will be inactive for the remainder of the procedure.

This follows immediately for Lemma 5. □

Observation 2

Once it is known that two neighbouring vertices x and y point to nodes in the same tree, then the edge processor (x,y) will no longer be needed.

This follows because edge processors are only used for hooking roots to other nodes, and since cycles are never created (Lemma 3), roots never hook to nodes within the same tree. □

As a corollary to Observation 2 we have the following.

Observation 3

If $\text{parent}[x] = \text{parent}[y]$ then (x,y) will thereafter be inactive.

Observation 4

If (x,y) succeeds in hooking the root parent[x] onto the vertex parent[y] in either Step 2 or 3 then the processor (x,y) will be inactive thereafter.

Connected component labelling 113

Given the inclusion of Step 4, we can identify yet another situation when processors will be inactive:

Observation 5

If, during Step 4, parent[u] is set to point to a root then the vertex u need not be considered in the subsequent Step 1.

Of course, shortcutting is unnecessary if u points to a root. However, it takes as much time to check whether u points to a root as it does to execute the shortcutting step for u. Thus this observation would not seem to help much. But if a vertex points to a root at the start of Step 4, then it will be among those vertices that point to roots after Step 4, and this information can be obtained essentially for free (i.e., parent[u] does not change in Step 4). These nodes need not be considered in the following Step 1.

Note that, owing to the possibility of hooking in Steps 2 and 3, all vertices except those in dead trees must be subjected to shortcutting in Step 4.

We see that for the most part, tasks to be performed by steps in the algorithm are pruned as processors become inactive. The only exception is that shortcutting a node in Step 1 can be omitted if the node is known to point to a root on that iteration; it may happen that the root will subsequently be hooked and so the processor will be reactivated. Nonetheless, we might be tempted to maintain a list of active processors, and remove items from the list as processors die according to the observations above. However, for purposes of synchronization, we need to form a new list for each step, and to read from the current list when assigning tasks in the current step. Thus it is easier to enqueue tasks that will be necessary in subsequent steps as they are discovered in the current step. We discuss the enqueuing and processor assignment methods in the implementation section.

How should task requests be represented when enqueued on a list? For Steps 1 and 4, tasks will be represented as a list of vertices to be subjected to a shortcutting step. For Steps 2 and 3, tasks are given as a list of edges between vertices. Strictly speaking, the edges are directed. However, if (x,y) is inactive, then (y,x) is also inactive, since our only criterion for inactivating edge processors is Observation 2, which is symmetric in x and y. So the entry (x,y) can stand for two edge tasks: namely (x,y) and (y,x). Further, Steps 2 and 3 can be recoded to perform both of the directed tasks at once as a single task. Initially, all vertices in all components are on the vertex list, and the edge list consists of all edges (x,y) with y as a south or east neighbour of x, and x and y both are vertices in the vertex list. The initial lists can be formed by a raster scan of the image, which can be done in parallel. (See the pseudocode for Step 0 in the Appendix.) Processors are assigned to successive pixels in raster-scan order. A processor assigned to pixel x enqueues the south and/or east edges (x,y) if the edge connects vertices in regions.

Step 1 is performed for all vertices on the vertex list. No new lists are formed during Step 1.

For Step 2 we examine all edges on the edge list. During Step 2, (x,y) attempts to hook parent[x] onto parent[y] or vice versa, under the conditions that one node is a root and the other has a larger numerical value. A new edge list is formed, and (x,y) is enqueued on that list unless (1) it is observed that parent[x] = parent[y] when checking the order relationship on the parents, or (2) the task (x,y) is successful in hooking parent[x] onto parent[y], or (3) the task (y,x), which is performed as part of the (x,y) request, is successful in hooking parent[y] onto parent[x]. It appears as though a slight bit of additional work is needed to perform these checks: ordinarily, parent[y] does not have to be accessed by (x,y) if parent[x] is not a root. The objection is moot, however, since the symmetric (y,x) operation must check parent[y] to see if it is a root, so that both parents will be accessed anyway. There is, nonetheless, a slight time penalty to determine whether a request to hook is successful, and also to perform the enqueue step, as we will discuss in the implementation section.

Step 3 then uses the new edge list, and can form yet another edge list. If a new list is formed, the process is identical with the edge list formation in Step 2: for each edge (x,y), parent[x] and parent[y] will both be accessed (to see if either is a stagnant root). The edge (x,y) is not enqueued if parent[x] = parent[y], or if parent[x] succeeds in hooking onto parent[y], or vice versa.

It is not clear that forming a new list (which will be used for Step 2 in the next iteration) is worth the extra expense during Step 3. Instead, one can simply retain the edge list formed during Step 2, and use it for the next iteration's Step 2. Some edges will not be pruned that are otherwise unnecessary, but they are likely to be rare, and will be pruned during the list formation in the subsequent Step 2. Empirical studies are needed to judge whether creating a new list in Step 3 is justified.

Finally, Step 4 uses the vertex list, and creates two new vertex lists. One vertex list will be used for the subsequent Step 1, and the other vertex list will be held for Step 4 of the next iteration. The Step 1 list is formed of all vertices that change parents during the Step 4 shortcutting. The remaining vertices are known to point to roots, and so need not be considered in the subsequent Step 1. The Step 4 list is formed of all vertices except those in dead trees. A member of a dead tree can be recognized in Step 4 by noting whether the vertex points to a root which has not been shortcut or hooked in

Connected component labelling

Steps 1, 2, or 3. That is, the vertex v is enqueued on the next Step 4 list unless parent[v] does not change and age[parent[v]] $< I$ in Step 4. Pseudocode for the modified steps is presented in the Appendix.

6 SYNCHRONIZATION

In a processor-poor environment we cannot execute all of the necessary tasks within a step at once. Since we wish to avoid copying the entire pointer graph and age labels to a separate output graph in each step, we must either queue the requests for changes to the pointer graph, and execute the changes after completing the analysis for the step, or suffer the consequences of asynchronous operation of each step. Since we have a list of tasks to be performed, we could enqueue the results of those tasks to a separate list, being sure to request a hook on a root no more than once (the first request to a node should block all further requests), and storing with all shortcut requests the node to which the new parent should point. The request list could be performed after the task list is exhausted (synchronizing between scans on the lists). It turns out, however, that in the case of the Shiloach/Vishkin connected-components algorithm, it is sufficient to operate each step asynchronously.

Within Steps 1 and 4, if shortcut requests operate asynchronously, trees will still reduce in height. The final tree may be different than the tree resulting from a synchronous shortcut step. For example, a four-node linear tree of height 3, if shortcut by 4 separate (sequential) tasks, starting from the root and propagating to the leaf, will result in a rooted star. The same tree, when shortcut by a parallel operation, results in a tree of height 2. However, in this example and all other examples, asynchronous shortcutting can only lead to more height reduction as compared to synchronous shortcutting. This is because a node's parent's pointer may have already been shortcut, in which case it can only point closer to the root. Thus the shortcut on the current node will yield a parent pointer that goes close to the root. Unfortunately, an improvement cannot be guaranteed, since it may happen shortcut tasks occur from the leaf nodes toward the roots, rather than the other way around.

If hooking operates asynchronously, the first time a root u is hooked to a node v, it ceases to be a root, and so will receive no more hook requests during that step (or, for that matter, in subsequent steps). However, the new edge from u to v might be responsible for causing another node u' to attempt to hook to v during the same step. Alternatively, if the node v happens to be a root, it might happen that it will attempt to hook onto a node u' on account of the new edge from u to v during that step. In either case, the situation arises if u has some neighbour x such that x points to u'. However, an analysis of the proof of the time bound shows that the only requirement of Steps 2 and 3 is that all eligible roots going in to a hooking step are in fact hooked. Asynchronous operation does not defeat this property, since asynchronous hooks only add possibilities for new hooks. None of the new hooks are harmful, since they only connect trees that comprise nodes in the same connected component. Some of the hooks may speed up the algorithm, by combining trees that otherwise would not be combined until later iterations.

We conclude that asynchronous parallel processing of the steps will at worst speed up the algorithm in terms of numbers of iterations.

The algorithm as given requires synchronization between the steps. Synchronization between Steps 2 and 3 is essential to properly define stagnant roots, and to prevent cycles. Synchronization between Steps 3 and 4 is needed to define dead trees, and to ensure that Step 4 does the useful work of reducing the height of combined trees. That is, if some shortcutting steps of Step 4 were to precede stagnant node hooking in Step 3, shortcutting might be attempted on a stagnant tree which is later hooked, and thus the combined tree height is not reduced by the requisite factor of $3/2$. Of course, the tree height would be reduced in the next Step 1, and so eliminating synchronization between these two steps is at worst equivalent to eliminating Step 4 entirely, which simply slows the algorithm by a multiplicative factor. Between Steps 4 and 1, synchronization is needed to properly test for termination, and again to ensure that Step 4 accomplishes all the work credited to it. The iteration counter is increased in a serial section between these two steps, and the test for termination is also done by a single serial processor. However, we could make the iteration counter a local variable, updated by each processor, and observe vertices not pointing to roots in Step 4, feeding them into processors which then execute the Step 1 shortcutting. This of course assumes a partly asynchronous operation of the steps, and means that a Step 1 shortcut on a node v will access v's parent, who may not yet have been shortcut by its Step 4 operation. The result is that Step 4 may be partly ineffectual, which as noted before at worst slows down the algorithm by a constant factor.

There is a minor penalty involved in requiring synchronization. Especially if the number of processors P is a large fraction of the number of vertices N, there will be a point at which all tasks for a step have been assigned and many completed, and available processors have nothing better to do than wait for completion of the remaining uncompleted tasks. Thus the degree of parallelism is hurt by the need to keep some processors idle.

Connected component labelling 117

7 IMPLEMENTATION

The Appendix contains pseudocode for performing the Shiloach/Vishkin algorithm on an MIMD machine. We have written the program assuming synchronization between all steps, although as noted in the previous section, synchronization is necessary only between Steps 2 and 3. We also assume asynchronous operation within each step.

In our C-like language, variables declared as "shared" are stored in the common memory, and so are accessible to all processors. The address of a shared variable (denoted by &v for the variable v) is also available to all processors. We have avoided declaring automatic or static shared pointers for clarity's sake; we could otherwise declare shared pointers (or variables in common memory containing addresses) pointing to shared variables, or local copies of addresses pointing to shared variables, or even perhaps pointers in common memory pointing to local variables, and it is not so obvious how the syntax should differentiate between these cases. In our algorithm, some of the procedure parameters are pointers to shared integers, and it is understood that the addresses of those shared integers are available to all processors within that procedure. That is, in

```
sub(pN)
shared int *pN;
{ . . .
}
```

all processors operating during { . . } have access to the integer *pN, and a copy of the address pN will also be located in common memory, but may well be cached by individual processors in their local memory if the processors handle the pointer as a read-only variable.

A block-structure language such as ADA or PASCAL might be more appropriate for coding these kinds of parallel algorithms, since the scoping rules for nested blocks would make the visibility of variables explicit.

We have used the "fetch-and-add" instruction as our principal coordination primitive. This instruction accesses a shared variable, returns the value, and increments the stored amount by a specified increment. That is, the instruction is equivalent to

```
fetch&add(pV, inc)
int *pV;
int inc;
{
    int val
    val = *pV
    *pV = val + inc
    return(val)
}
```

The important point is that concurrent fetch-and-adds to the same shared variable behave as though they had been requested serially, in some arbitrary order. Thus ten simultaneous "fetch&add(&v,1)" requests will increment v by 10, and return ten distinct consecutive integers, one to each calling processor. An important property of the Ultracomputer is that simultaneous fetch-and-adds to a single variable incur no time penalty. That is, all processors can issue a fetch-and-add to the same variable in the same amount of time as it would take for one processor to execute a fetch-and-add.

We use the construct "cobegin . . . coend" to allocate processors. The "cobegin nomorethan N" line denotes a request to the operating system to assign as many processors as are available, but no more than the integer value in the variable N, and put those processors to work executing the following code. The processors use the "fetch-and-add" primitive to dynamically allocate unique identifiers, which leads to data variability and hence execution variability amongst the processors. The multiple processors thus execute the intervening code in an asynchronous fashion. The "coend" line is interpreted as requesting that all but the last assigned processor reaching the "coend" statement be yielded back to the operating system, and perhaps put into an idle loop. The last processor to reach the "coend" statement proceeds to execute the code following the "coend".

The operating system might implement the "cobegin . . . coend" construct by spawning exactly the requested number of processes upon reaching the cobegin, leaving it to the scheduler to share processors among requested processes. A shared variable is loaded with the number of active processes. As a process completes, by reaching the "coend" statement, a "fetch-and-add" instruction with an increment of -1 is issued to the shared variable representing the number of active processes. The last process will receive the value 1 from the fetch-and-add instruction, and thus will know not to die, but instead to continue by reading instructions following the "coend".

An alternative approach allows the "cobegin" to instantiate available processors to the intervening code, along with a shared variable which counts the number of assigned processors. As processors reach the "coend", they issue fetch-and-adds with an increment of -1. Once again, the last processor will receive a value of 1, and thus know to continue with the remaining code.

Within each "cobegin . . . coend" block, we consider each step as a set of independent and potentially parallel subtasks. The subtasks are given by the processing of each vertex or edge. We organize the vertices and edges into lists, which are represented by arrays indexed by a shared variable. In the

Connected component labelling

fully parallel case, if a sufficient number of processors are available, we use a processor for each vertex or edge element on the list. A processor requests a new subtask by executing a fetch-and-add to the shared index with an increment of 1. The value returned is a pointer to the element (edge or vertex) that is to be processed in the subtask. The nature of the fetch-and-add instruction ensures that no two processors will receive the same element. The guarding "while" loop ensures that all tasks are assigned before any processor is yielded to the "coend" statement, and that any available processor requests the next unassigned task.

Note that if the allocation of processors is done with a fixed unique process identification number for each, then that number can be used to coordinate the allocation of tasks so that no two processors are assigned the same task. However, the dynamic allocation of tasks becomes much more difficult without a fetch-and-add instruction. The time penalty for synchronization between steps is likely to be much greater if the task allocation is based on static processor identification numbers.

In our code for the Shiloach/Vishkin connected-component algorithm, we assume that the image is a globally available raster-scan array of binary values (1s and 0s). We construct the pointer graph, which will also be global. We structure the pointer graph as an image of indices in registration with the input image: at a given pixel, the value of "Parent(i)" at that pixel is the index of the pixel pointed to by that node in the pointer graph. This structure makes the code more clear, but wastes a lot of space, since Parent(i) is undefined for pixels i that are not within regions in the binary image (i.e. where Img(i) = 0). Instead, the pointer graph should be a set of nodes, with each node containing fields specifying the coordinates of the corresponding pixel, the Age field, and a pointer to the parent node in the pointer graph. Since our pointer graph is in registration with the image, and indexed by the pixel coordinate, the Age field is also a registered image. We also use a "Hooked" tag on each node, which is used to determine whether a node is successful in hooking onto another node in Steps 2 and 3.

The vertex and edge lists are simply arrays of indices and pairs of indices respectively. We use the fetch-and-add primitive to enqueue elements onto the lists. Initially, all vertices within regions in the image are placed on the vertex list, and all south and east edges in regions are placed on the edge lists, as built up in Step 0. Step 1 shortcutting is unnecessary during the first iteration, and so is skipped. In all of the steps, a fetch-and-add is issued to an index in order to allocate elements on the list to processors. In Steps 2 and 3 the current edge list is scanned, and a new edge list is formed. Step 2 creates Step 3's edge list, while Step 3 creates the edge list for the next Step 2. Each processor enqueues the edge element it is currently processing onto the output list, unless it observes that both vertices of the edge point to the same node, or unless the current edge is successful in causing a hook. Since several processors may request a hook of a single node simultaneously, it is necessary for a processor to determine whether it has been successful in causing a hook. This is done by means of a hook count attached to each node, represented as the array "Hooked(i)". In order for a processor to gain permission to hook node i, it must issue a fetch-and-add with increment 1 to the variable Hooked(i), and retrieve the value 0. Since only one processor will receive a 0-value from the fetch-and-add, exactly one hook will take place and a processor knows whether it is the winner according to the permission granted from the returned value. In this scheme, we are using the fact that once a node hooks onto another node, then it is no longer a root and will at no future time be eligible for hooking again.

During Step 0 the vertex list for Step 4 is created, and stored in the array Vlist4a. The vertex list for Step 1 is not needed, since Step 1 is skipped during the first iteration. During Step4 the Vlist4a list is read, and new lists of vertices are created, one for Step 1 and one for the next iteration's Step 4. The new list for Step 4 is written to Vlist4b. If we instead wrote to Vlist4a, overwriting the list being scanned, we could potentially overwrite a value in the list that has been allocated to a processor but not yet read. The length of the list, NV4, is continually overwritten, but the length of the original list is passed by value and stored in the procedure-local variable NV. In alternate iterations we then read from Vlist4b and write to Vlist4a. Finally, note that Age fields are updated after each shortcut and hook operation, updating the target of the shortcut or hook in Steps 1, 2 and 3. A dead tree is one whose root is stagnant after Step 3. Note that a hook into a stagnant tree must hook onto the root, so that if a stagnant tree is made live by Steps 2 or 3 the Age field of the root is updated. Thus in Step 4, nodes in dead trees are recognized because they point to roots whose Age field has not been updated in the current iteration.

Our time bound of Section 3 shows that the algorithm will terminate after no more than $\log_{3/2} N$, i.e. approximately $1.71 \log_2 N$ iterations. Inspecting the algorithm, we see that Step 1 requires 5 accesses to the common memory, while Steps 2, 3 and 4 require 11 accesses per task. The time used for Step 0 is negligible. In counting the number of accesses, we have assumed that a fetch-and-add is equivalent to two accesses, that independent reads or writes can be pipelined through a connection network and thus can be performed essentially concurrently, and that accesses by separate processors do not interfere. These are all reasonable assumptions, modulo minor perturbations, for the Ultracomputer.

The unknown variable in an analysis of the MIMD approach to the Shiloach/Vishkin algorithm is the rate of decrease of the task list sizes. Initially, the number of tasks are $NV1 = NV4 = N$ and $NE2, NE3 < 2N$. We

Connected component labelling

conjecture that the edge lists will normally drop off geometrically. The vertex lists, particularly the list for Step 4, will drop only after a component is completely labelled, which can take 1.7 log N_c iterations, where N_c is the size of the component. In any case, no iteration will need more than $60N$ accesses, and considerable drop-off is expected as iterations proceed.

Let N_c be the number of nodes in the maximum size connected component and N the total number of nodes. Suppose that there are P processors, and each access to shared memory takes $\alpha \log P$ time (in seconds). Since P processors can be devoted to performing tasks simultaneously, we come up with a time bound of

$$\frac{103\alpha N \log N_c}{P/\log P}$$

seconds for termination of the algorithm. If we dispense entirely with the enqueuing of new tasks, using the same vertex and edge lists in all iterations, we can then modify the code so that 5 accesses per task are needed in Step 1, and 6 accesses per task suffice in Steps 2, 3 and 4. In this case, we know that roughly 35 N accesses will be needed per iteration, with no drop-off. We can thus drop the 103 constant in our upper bound to 60, for otherwise the enqueuing process is not supportable. The expected constant is probably somewhat smaller, due to the drop in the length of the task lists.

We can also inspect the code to determine what percentage is spent in allocation and enqueuing of tasks, and what percentage is spent in actually performing the steps of the algorithm. This gives a measure of the degree of parallelization. The measure is still somewhat misleading, however, since the use of an interconnection network or other access strategy exists solely to support concurrent reads and writes, and is thus an additional price paid for parallelism. The synchronization between steps introduces an additional drop in the degree of parallelization, which is difficult to measure since it depends upon the load on the parallel computer and the number of available processors relative to the size of the task lists. Finally, many of the tasks attempt to perform a shortcut or hook and are unsuccessful in the attempt, either because of multiple requests or because the conditions are not satisfied to request the action. These tasks might be considered wasted effort. Just taking into account allocation and enqueuing, we estimate that nearly 50% of the accesses during shortcutting steps and only 30% during hooking steps constitute performance of the algorithm. The rest is overhead for parallelization.

Typical numbers for a 512 by 512 image would be $\alpha = 25$ ns, $N = 512^2/8$, $N_c = 4000$ and $P = 4096$. This yields a time bound (using the constant 60) of 1.728 ms.

This compares with the standard sequential methods, which permit two frame-time (67 ms) connected component labellings on 512 by 512 images. By using a supercomputer with a 10 ns cycle time, a linear connected component can be performed in something like $10n^2$ cycles, which for $n = 512$ yields 26 ms times.

Let us briefly compare these times with an n by n mesh connected parallel array of single-bit processors. Let us assume the breadth-first-search method instead of the Nassimi/Sahni algorithm. Each pixel requires a unique processor identifier, which for an n by n mesh will have $2 \log n$ bits. A local max among the central pixel and the four nearest neighbours can be constructed using a sequence of 5 pairwise maxima. The maximum of two integers a and b can be calculated using the formula max $(a,b) = \frac{1}{2}(a+b+|a-b|)$. Thus roughly 10 arithmetic operations, each involving integers with $2 \log n$ bits, are required per iteration. In addition, a $2 \log n$ bit operation is needed to mask the boundary, and another operation is needed on each iteration to check for termination. We assume the existence of a global "sum-or" which can detect a non-zero value within a designated plane in the grid in one cycle time. Combining, we estimate that $24 \log n + 1$ cycles are needed per iteration. How many iterations will be needed, on the average? We are tempted to answer $N^{\frac{1}{2}}$ as an estimate of the diameter of the largest component. Unfortunately, the components are frequently boundaries of connected regions, and a safer bet for the maximum internal path length is $\frac{1}{2}n$. Thus we arrive at the time bound of

$$\alpha n \ (12 \log n + \tfrac{1}{2}).$$

Using $n = 512$ and $\alpha = 50$ ns, which seem to be currently typical numbers, we obtain a time bound of 2.78 ms.

We see that for these numbers, the MIMD Shiloach/Vishkin approach wins, although the bounds are within the same order of magnitude. Thus the true benefits of MIMD approaches to image processing lie in the greater degree of flexibility, and performance improvements, if present, will be borne out mostly by realistic empirical studies.

ACKNOWLEDGMENTS

Work on this paper has been supported in part by NSF Grant DCR-8403300. An earlier draft and much of the research was conducted by Alan Rojer, whose assistance is greatly appreciated. He was supported by Navy Grant N00014-85-M-0260.

REFERENCES

[1] Duff, M. J. B. (1978). Review of the CLIP image processing system. In *Proc. Nat. Computer Conf.* p. 1055. See also Duff, M. J. B. (1985). Real applications on CLIP4. In *Integrated Technology for Parallel Image Processing* (ed. S. Levialdi), 153–165. Academic Press, London.
[2] Batcher, K. E. (1980) Design of a massively parallel processor. *IEEE Trans. Comp.* **29,** 836–840.
[3] Kung, H.T. (1982). Why systolic architectures? *Computer* **15,** 37–46.
[4] Hummel, R. A. (1984). Image processing on the NYU Ultracomputer. *Courant Inst. NYU Ultracomputer Note* No. 72.
[5] Nassimi, D. and Sahni, S. (1980). Finding connected components and connected ones on a mesh-connected parallel computer. *SIAM J. Comp.* **9,** 744–757.
[6] Miller, R. and Stout, Q.F. (1984). The pyramid computer for image processing. *Proc. 7th Int. Conf. on Pattern Recognition.* pp. 240–242.
[7] Gajski, D., Kuck, D., Lawrie, D. and Sameh, A. (1983). CEDAR—a large scale multiprocessor. In *Proc. Int. Conf. on Parallel Processing,* p. 524.
[8] Gottlieb, A., Grishman, R., Kruskal, C.P., McAuliffe, K.P., Rudolph, L., and Snir, M. (1983). The NYU Ultracomputer—designing an MIMD shared memory parallel computer. *IEEE Trans. Comp.* **32,** 175–189.
[9] Siegel, H. J., Siegel, L. J., Kemmerer, F. C., Mueller, P. T., Smalley, H. E. and Smith, S. D. (1981). PASM: a partitionable SIMD/MIMD system for image processing and pattern recognition. *IEEE Trans. Comp.* **30,** 934–947.
[10] Reeves, A. P. (1985). Multicluster: an MIMD system for computer vision. In *Integrated Technology for Parallel Image Processing* (ed. S. Levialdi), pp. 39–56. Academic Press, London.
[11] Shiloach, Y. and Vishkin U. (1982). An $O(\log n)$ parallel connectivity algorithm. *J. Algorithms* **3,** 57–67.

APPENDIX. MIMD SHILOACH/VISHKIN CONNECTED COMPONENTS PSEUDOCODE

```
/*  Shiloach/Vishkin algorithm, coded in pseudo-C code for
 *  the NYU Ultracomputer
 */

    #define   n = 512         /* Image is n by n */
    typedef   pnode = integer range[0..n² - 1];
    typedef   edge = record
              {x:  pnode;
               y:  pnode;}
    /*globals:*/
    shared    binary   Img(0..n² - 1); /* Input image is n by n, raster scan order */
    shared    pnode    Parent(0..n² - 1);
    shared    int      Age(1...n²), Hooked(1..n²);
```

```
procedure Shiloach Vishkin()
{
    shared    pnode    Vlist1(0..n² – 1), Vlist4(0..n² – 1);    /* Vertex lists */
    shared    edge     Elist2(1..2n²), Elist3(1..2n²);          /* Edge lists */
    shared    int      NV1, NV4, NE2, NE3;  /* Integers giving the length of lists */
    shared    int      I;                   /* Iteration counter */

    /* begin */
    step0(Vlist4a,&NV4,Elist2,&NE2);   /* Create vertex list for step 4 and Edge list for step 2 */
    I←0;
    while (NV4 > 0)                    /* While there are non-dead trees */
    {I←I + 1;
        if (I > 1)step1(Vlist1,NV1);           /* Uses Vertex list Vlist1 */
        step2(Elist2,NE2, Elist3,&NE3);        /* Using Elist2, create Elist3 */
        step3(Elist3,NE3, Elist4,&NE2);        /* Using Elist3, create Elist2 */

        if (isodd(I))
            {step4(Vlist4a,NV4,Vlist1,&NV1,Vlist4b,&NV4);}
        else
            {step4(Vlist4b,NV4,Vlist1,&NV1,Vlist4a,&NV4);}
    }
}


step0(Vlist,pNV,Elist,pNE)    /* Initializes pointer graph and loads Vertex list and Edge list */
    shared    pnode    Vlist(0..n² – 1);
    shared    edge     Elist(1..2n²);
    shared    int      *pNV, *pNE;
{
    shared    pnode    pI;

    pI←0; *pNV←0; *pNE←0;
    cobegin nomorethan n² – n
        private    pnode    i;
        private    int      k;
        {while ((i←fetch&add- (&pI,1)) < n² – n)

                                        /* We assume that the bottom row and
                                         * right column of Img are all 0's */
            {if (Img(i) = 1)
                {Parent(i)←i;                       /*Self pointing root */
                Age(i)←0;
                Hooked(i)←0;                        /* Counter for hook requests */
                k←fetch&add(pNV,1);                 /* Enqueue node i */
                Vlist(k)←i;
                if (Img(i + 1) = 1)                 /* Enqueue east neighbor */
                    {k←fetch&add(pNE,1);
                    Elist(k).x←i;
                    Elist(k).y←i + 1;}
```

Connected component labelling

```
                        if (Img(i + n) = 1)              /* Enqueue south neighbor */
                            {k←fetch&add(pNE,1);
                            Elist(k).x←i;
                            Elist(k).y←i + n;}
                        }
                 }
         }
coend
}

step1(Vlist,NV)         /* Shortcuts all nodes on Vlist */
shared   pnode       Vlist(0..n² – 1);
shared   int         NV;
{
shared   int      index;
index←0;
cobegin nomorethan NV
    private   pnode       old-parent, new-parent;
    private   int    k;
    {while ((k←fetch&add(index,1)) < NV)
        {old-parent←Parent(k);
         new-parent←Parent(old-parent);
         Parent(k)←new-parent;
         if (old-parent ≠ new parent) Age(new-parent)←1;
        }
    }
coend
}


step2(Elist,NE,Elistout,pNEout)
shared   edge     Elist(1..2n²);
shared   int      NE;
shared   edge     Elistout(1..2n²);
shared   int      *pNEout;
{
shared   int      index;

index←0
*pNEout←0;
cobegin nomorethan NE
    private   edge      e;
    private   pnode     u,v;
    private   int   k,h;
    {while ((k←fetch&add(index,1)) < NE)
        {h←1;
         e←Elist(k);                    /* Edge is (e.x,e.y) */
         u←Parent(e.x);
         v←Parent(e.y);
         if (u < v&&Parent(u) = u)
             {h←fetch&add(&Hooked(u),1);
```

```
                    if (h = 0)               /* Allow hook only if not hooked yet */
                        {Parent(u)←v;
                         Age(v)←I;
                        }
                    }
                else if (u > v && Parent(v) = v)
                    {h←fetch&add(&Hooked(v),1);
                     if (h = 0)               /* Allow hook only if not hooked yet */
                        {Parent(v)←u;
                         Age(u)←I;
                        }
                    }
                if (u≠v && h≠0)
                    {k←fetch&add(pNEout,1);
                     Elistout(k)←e;
                    }
                }
            }
    coend
}

step3(Elist,NE,Elistout,pNEout)
shared   edge   Elist(1..2n²);
shared   int    NE;
shared   edge   Elistout(1..2n²);
shared   int    *pNEout;
{
shared   int    index;

index←0;
*pNEout←0;
cobegin nomorethan NE
    private   edge    e;
    private   pnode   u,v;
    private   int     k,h;

    {while ((k←fetch&add(index,1)) < NE)
        {h←1;                               /* Hooked only if h = 0 */
         e←Elist(k);
         u←Parent(e.x);
         v←Parent(e.y);
         if (u≠v)
            {if (Age(u) < I && Parent(u) = u)
                {h←fetch&add(&Hooked(u),1);
                 if (h = 0)               /* Allow hook only if not hooked yet */
                    {Parent(u)←v; /   * Hook u to v */
                     Age(v)←I;}
                }
             else if (Age(v) < I && Parent(v) = v)
                {h←fetch&add(&Hooked(v),1);
                 if (h = 0)               /* Allow hook only if not hooked yet */
```

Connected component labelling

```
                    {Parent(v)←u; /* Hook v to u */
                     Age(u)←I;}
                    }
             if (h≠0)
                    {k←fetch&add(pNEout,l);
                     Elistout(k)←e;
                    }
                   }
            }
      }
coend
}

step4(Vlist,NV,Vlist1,pNV1,Vlist4,pNV4)
shared    pnode    Vlist(0..n² – 1),Vlist1(0..n² – 1),Vlist4(0..n² – 1);
shared    int      NV;
shared    int      *pNV1,*pNV4;
{
shared    int      index;
#define   NOT      !

index←0;
pNV1←0; *pNV4←0;
cobegin nomorethan NV
       private    pnode    old-parent, new-parent;
       private    int      k,j;

       {while ((k←fetch&add(index,1)) < NV)
              {old_parent←Parent(k);
               new_parent←Parent(old_parent);
               Parent(k)←new_parent;
               if (old_parent ≠ new_parent)
                    {j←fetch&add(pNV1,1);
                     Vlist1(j)←k;}
                 if NOT(old_parent = new_parent && Age(new_parent) < I)
                    {j←fetch&add(pNV4,1);
                     Vlist4(j)←k;}
               }
       }
coend
}
```

PYRAMIDS

Pyramid architectures have been widely acclaimed as candidates for tackling the problem of intermediate-level processing. The base of the pyramid can be treated as a conventional SIMD rectangular array of processors, but the higher layers, in some way or another, are used to data emerging from the lowest layer and therefore, by definition, are performing intermediate-level processing.

Leonard Uhr reviews the potential of some of the more complex pyramid structures that have been suggested and also proposes some further classes of pyramid which should be well suited to particular types of algorithm. Quentin Stout analyses various mesh-structure computers as well as pyramids and derives algebraic expressions for the theoretical performance of the structures when implementing algorithms in certain classes which he defines, paying particular attention to the performance restrictions imposed by interprocessor communication. David Schaefer examines three distinct pyramid designs now being constructed, and uses a logic notation he devised for SIMD arrays in order to describe and compare the three systems. This section closes with a description by Virginio Cantoni and Stefano Levialdi of an algorithm for contour labelling which indicates the locations of convexities and concavities. Although the algorithm is ideally suited to SIMD arrays, it can usefully be extended into a pyramid implementation in order to overcome the difficulty of specifying the necessary size of the processing neighbourhood for a range of curvatures in the labelled contour.

Chapter Eight
Multiple-Image and Multimodal Augmented Pyramid Computers

L. Uhr

1 INTRODUCTION

There has been a good deal of interest during the past few years in developing massively parallel hierarchical converging pyramid structures (see, e.g. [1–11]), both hardware and software, for image processing, pattern recognition and computer vision. Pyramids, when programmed in appropriate ways that take advantage of their architecture, appear to have the potential of great speed and efficiency in executing a wide variety of perceptual operations.

They have the massively parallel structure of arrays, with all the potentials for speed-ups when executing the parallel local operations of which arrays are capable. In addition, with only minor increases in hardware (at most one third additional computers) they are able to execute global operations in $O(\log N)$ time that on an array would take $O(N)$ time. Also, when successive images are pipelined through the layers of the pyramid, an additional $O(\log N)$ speed-up can be achieved. Possibly most important of all, pyramids have an overall (massively) parallel-(shallowly) serial converging/diverging hierarchical structure that appears to be especially appropriate for brain-like processing of images [9, 12].

A pyramid has the following overall structure: successively smaller arrays (alternately called grids or meshes) of computers are linked together as though by a tree. Thus an $(N+1)$-dimensional pyramid can be thought of as a linked stack of N-dimensional arrays. Alternatively, it can be seen as a base "retinal" array linked to a tree that has been augmented with extra links that make possible lateral array-like processes within each layer, as well as converging and diverging tree-like processes moving up and down from layer to layer.

1.1 The tree-like structure linking stacked arrays into a pyramid

Pyramid structures have chiefly been used to handle two-dimensional inputs, of the sort produced by television and cameras. Therefore the hardware pyramids designed or under construction have been three-dimensional structures with two-dimensional arrays at the base. Each array layer is $D \times D$ times as large as the "parent" layer "above" it (that is, toward the pyramid's apex). Since D (divergence/convergence) should be an integer, the smallest symmetric increase in size is $2 \times 2 = 4$. This structure gives a logarithmic reduction in size, and a logarithmic number of layers. An $N \times N$-based pyramid with $D \times D$ divergence/convergence will have $1 + \log_D N$ layers.

Most (but not all) pyramid designs specify that each layer is $2 \times 2 = 4$ times as large as its parent layer. Thus, for example, a pyramid with a 256×256 base array and a 2×2 divergence/convergence will have 9 layers.

Links from one layer to the next are each to their nearest computers. For example, consider an $N \times N$-based pyramid with 2×2 divergence/convergence. The pyramid's apex (a 1×1 array with only one computer, called layer 0) links to four "child" computers arranged in (and often physically linked into) a 2×2 array, forming layer 1. Each of these four computers links as a parent to four children in a 2×2 subarray of layer 2, these four 2×2 subarrays tiling out into a 4×4 array. Each of these sixteen computers has four children in layer 3, forming an 8×8 array. This structure continues until the $N \times N$ array at layer $\log N$ is reached.

1.2 The structure of the single array, and the processes it executes

An image is input to the large array that forms the base of the pyramid. This array (as are all the other arrays in the pyramid) is, essentially, like the two-dimensional arrays that were first proposed by Unger [13], partially built by McCormick [14] (32×32 Illiac-III), and by Slotnick, Borck and McReynolds [15] (32×32 Solomon-I), and built and commissioned by Duff [16] (96×96 CLIP4), Reddaway [17] (64×64 DAP) and Batcher [18] (128×128 MPP). This kind of array links each of its thousands of computers to its nearest neighbours, in grid form. (All the arrays actually built, with the exception of CLIP, link each computer to its 4 square neighbours—north, east, south and west. CLIP links each computer to its 8 square and diagonal neighbours—the above plus northeast, southeast, southwest, northwest. CLIP is also reconfigurable, under program control, to link each computer to its 6 hexagonal neighbours.)

The CLIP, DAP and MPP arrays have a single controller. Therefore the same sequence of instructions is being executed everywhere, by all the array's computers, in parallel in SIMD ("single instruction, multiple data-stream" [19]) fashion. Their computers also have very simple 1-bit processors (but it is important to emphasize that these are general-purpose—they carry out N-bit operations, e.g. 32-bit arithmetic, bit-serially). And the memory of each computer is quite small compared with memories in conventional computers, ranging from 32 bits (CLIP4) to 16 Kbits (DAP)—although the memory of the entire multicomputer can be quite large (e.g. DAP has 8 MBytes). It is important to emphasize that the single controller, 1-bit processors and small memories are not inherent in or necessary to the array design. They are design choices based almost entirely on the great expense that would be entailed in building arrays with thousands of 32-bit processors, each with its own controller and a conventional memory (e.g. 1 MByte).

These arrays, and all of a pyramid's arrays, execute typical near-neighbour array operations of the following sort: information is fetched by each and every computer from a specified memory location in that computer and also in the four, eight or six directly linked computers. A general-purpose processor computes whatever logic, arithmetic, or matching operation is designated. The results are stored, for each and every computer, in specified locations in that computer's memory. (The fact that all this is done simultaneously for each and every computer in the array is what gives the array its massive parallelism.) Processes that compute functions of information not immediately accessible over direct links to nearest-neighbour computers must shift that information, step by step, until it is.

2 PYRAMID OPERATIONS

In addition, each computer in a pyramid can fetch and process information from the children and parent computers to which it is directly linked. Successive transformations can be effected both within a layer and from layer to layer, giving successively more abstract information about images. This means that information can be transformed, moved, converged and combined up the pyramid, and also diverged and broadcast down the pyramid.

2.1 Designs for pyramids, and for overlapped and truncated pyramids

Several researchers (e.g. [1,6,7,11,20,21]) have examined how hardware pyramids might be used, and have proposed preliminary designs. Three pyramids are now at preliminary stages of construction. Tanimoto (personal communication) has designed and partially checked out a pyramid chip using the NSF–DARPA silicon foundry facility. Cantoni et al. [22] have designed and are checking out a chip that they plan to use in a small prototype pyramid. Schaefer [23] is building a prototype 16 by 16-based pyramid using MPP chips (a 4 by 4-based system has been running since May 1985). Each computer links directly to its 4 square sibling neighbours (Tanimoto's link to all 8 sibling neighbours), plus 4 children and 1 parent.

The individual computers in these pyramids are all designed with the 1-bit processors and relatively small memories that are today typical for arrays. But they are all (with the possible exception of Tanimoto's) designed to have different instructions executed at each layer.

A number of alternative structures are possible [24,25].

For example, computers might be linked only to children and parents. If each links only to one parent, the result is a tree. Thus a "quad-tree" [26] is formed when each computer links to one parent and four children. A tree has the major flaw that adjacent pixels in the image often cannot be examined together except by moving very high up into the trees, possibly all the way to the root apex cell. But "overlapped" pyramids, where each computer is linked to several parents (a 2×2 of four parents is probably the minimum practicable number) overcome this problem. This kind of system can be designed so that layers of memory alternate with layers of processors, with each memory shared by child and parent processors. Now a parent can compute functions of its children, just as a computer in an array can compute functions of its siblings.

The pyramid might be "truncated", by chopping off the top few smaller layers. This is probably reasonable only when the pyramid is combined with a conventional computer or a network of asynchronous computers to handle the highest levels of processing.

A linear, rather than a logarithmic, pyramid can be constructed by taking a three-dimensional array and eliminating 1, 2, 3, . . . , $\frac{1}{2}N$ border cells at each layer moving up into the third dimension. This gives a pyramid whose height is $\frac{1}{2}N$ rather than log N. Linear pyramids probably need too much hardware and converge too slowly. But several layers from such a structure might be inserted between all or some layers of a logarithmic pyramid. Thus a pyramid of whatever depth is deemed desirable can be constructed, and not only the relatively shallow log N or the relatively deep $\frac{1}{2}N$.

The amount of convergence might vary from layer to layer. For example, the pyramid might converge rapidly near its very large base, but more slowly toward the apex.

Multiple-image and multimodal pyramids 135

3 DESIGNS FOR AUGMENTED PYRAMIDS

Several preliminary designs have been proposed for augmenting pyramids with additional hardware—usually a conventional serial computer, or an MIMD ("multiple instruction, multiple data-stream"—that is, asynchronous, each computer with its own controller) network of serial computers.

3.1 Augmenting the power of the hardware within the pyramid

The pyramid can be given successively more powerful processors moving upward [24]. For example, the base layer might have 1-bit processors, the next layers 2-bit, 4-bit, 8-bit, 16-bit, 32-bit and 64-bit. Alternately, the byte size might increase by powers of 4 rather than powers of 2, so that each processor can be configured to input from a square array.

Each of these processors might be made reconfigurable [27], to either 1 B-bit or B 1-bit processors, or even to in-between decompositions of B/D D-bit processors.

A $C \times C$ array of controllers can be assigned to an array, an $(N/C) \times (N/C)$ subarray of computers assigned to each.

Controllers can be assigned to arrays via a reconfiguring network [28,29], so that they can be assigned and reassigned, as desirable, under program control.

3.2 Augmenting the pyramid with a network of additional computers

The pyramid (either simple or augmented internally) can also be augmented by linking it to an external conventional computer or network of computers.

Pfeiffer [30] has designed hardware that will shift information out of the corners of a pyramid (or an array) and at the same time convert it into a chain code and store it into the memory of one or more conventional computer(s).

Tanimoto [31] is designing a pyramid that shares an $N \times N$ B-bit memory with a network of B computers, a different one of the network's computers linked to each $N \times N \times 1$ memory.

Ahuja [32,33] is exploring and suggesting designs of systems that reconfigure a number of trees and pyramid substructures into the larger architecture.

Uhr [34–36] has suggested a number of ways of augmenting pyramids, by increasing power within the pyramid itself (see above), or/and by linking the pyramid to appropriately structured networks.

Possibly the simplest such augmentation combines the pyramid with a conventional serial computer. For example, link a 16-bit computer to a scattered 4×4 array of 1-bit "connector" computers in the pyramid, each of which is at the centre of a 4×4 array in the 16×16 layer. Similarly, link a 64-bit computer to a scattered 8×8 array of connector computers each centred in a 4×4 array in the 32×32 layer. (Note that these will not be completely centered in the 4×4 array. So, for example, 2 instructions would be needed to shift information into these connector computers from any of the 16 computers in their 4×4 array. A 5×5 array would similarly need at most 2 instructions. So arrays whose dimensions were odd numbers might be preferable.)

It is probably preferable to use a whole MIMD network. Each B-bit computer might be linked to B pyramid computers, arranged in a suitable scattered or compact square or rectangular subarray, in the base array, or in any other appropriate layer.

Alternatively, the MIMD network might be linked to computers in several layers of the pyramid. Each computer in the MIMD network might be linked to several layers, or different computers might be linked to different layers (in which case information would have to be sent via several links within the MIMD network). This would allow information to be processed and then fed back from higher to lower levels of the pyramid or fed up from lower to higher levels.

Rather than linking the MIMD computers into an array, a variety of other topologies appear to be preferable. A pyramid of MIMD computers appears to be especially appropriate for linking to several, or to all, levels of the basic pyramid.

"Density", which refers to the number of nodes within a given "diameter" (the longest shortest distance between any pair of nodes) and "degree" (the number of links to each node), is widely considered desirable, to minimize message-passing distances. Arrays are quite poor with respect to density; pyramids are much better. But a large number of substantially denser graphs have been discovered in recent years (for a summary see [37]), including several that have been proved to be optimally dense. In particular the 10-node Petersen graph and the 50-node Singleton graph are optimal [38]. A variety of other graphs are also relatively much denser than those, like N-cubes and trees, that are more commonly used today for multi-computer networks [11]. These include constructions that compound good graphs with the complete graph (e.g. 51 Singleton graphs, each linked via a different node to each of the other 50 graphs) and augment trees by adding judiciously placed links (e.g. Goodman and Sequin's hypertrees [39] or de Bruijn graphs [40]).

Multiple-image and multimodal pyramids 137

4 PYRAMIDS EXTENDED TO HANDLE MULTIPLE IMAGES

The basic pyramid structure can be used to process a sequence of images, for example a continuing stream of frames from a TV camera—but serially, one at a time. It could also, in theory, be used to input binocular, I-ocular, or multimodal images one at a time. Once all these images have been input into the pyramid and stored in its computers' memories, the pyramid can combine, process and transform this information in any way that the programmer desires. This is so trivially, because a pyramid is a general-purpose system. More important, pyramids appear to be reasonably well structured to handle such processes. But there appear to be attractive extensions indicated to the basic pyramid design.

5 PYRAMIDS FOR BINOCULAR VISION

Possibly the simplest case is a 2-pyramid structure to be used to process binocular images. Such a system assumes that two input sensors (e.g. two television cameras or two arrays of photosensors) are focused from slightly different locations and at slightly different angles at what is usually the same region. They might be focused at exactly the same region, or overlapped to some known degree, or with space between the two different regions. A multicomputer structure for this problem can be designed in several ways.

The following assume that both images come from exactly the same region.

5.1 Two truncated or untruncated pyramids linking into a combining pyramid

Pyramids can be merged together as follows. Truncate two pyramids by removing layers, starting at the apex, so that both have top layers of the same size. Link the computers in these top layers to parents in the base of a third combining pyramid. This should probably be one of the pyramids just chopped off, but it might be some other pyramid. Link each computer in the base of the combining pyramid as a parent to eight child computers (four in each of the two lower-level pyramids, rather than the standard four in toto). Using some other combining pyramid whose base is a different size, link each parent in its base to the requisite number of children in the top layers of the two truncated pyramids.

A number of variants are now possible.

The combining pyramid might have more powerful processors. With 2-

bit processors, each processor could be joined to eight children by linking each bit to the four children in one of the lower-level pyramids. The processors could also, where indicated, be given additional hardware to execute bit-matching and bit-comparing operations between these two bits.

With 4-bit, 8-bit, 16-bit or 32-bit processors the number of processors in the higher-level pyramid could be reduced accordingly, and the input from the lower-level pyramid's computers effected at whatever speed desired.

Only one pyramid might be truncated. It would then output to the appropriate layer of the second, complete, pyramid—which would have additional wires and hardware to input this double amount of information.

5.2 Linking internally augmented pyramids and augmenting networks

Two pyramids could be used, each with successively more powerful processors moving upward. Now both might be truncated and linked to a suitably more powerful array in a similar higher-level pyramid, or one might be truncated and linked to the second at any desired layer.

Two truncated pyramids might both output into an augmenting network of any sort, for example an MIMD pyramid or tree, a Petersen or Singleton graph, or a network of any other desired topology.

5.3 Combining partially overlapped or non-overlapped images

To handle partially overlapped regions, the combining pyramid must input transformed images from the two lower-level pyramids into a rectangle that is a function of the combined region.

If the images overlap $V\%$ in the horizontal direction, the combining pyramid should be $N \times (N + N^*V)$. For example, if the images are combined at the 32×32 layer, where the overlap is $\frac{1}{2}$, or 16 cells, the combining pyramid should be 32×48.

If the images do not overlap at all, a rectangular hull drawn around them can be used to designate the combining array's base and the placement of the two images into that array. Consider the example where the transformed images are to be combined at the 32×32 layer, and they are 16 apart (that is, the distance between them in the original image is half the width of each of them). In this case the combining array should be 32×80. The first array should input to the 32×32 subarray at the extreme left; the second should input to the 32×32 subarray at the extreme right.

Multiple-image and multimodal pyramids 139

6 *I*-OCULAR PYRAMIDS

Rather than link two eyes, as for standard binocular vision, we might want to link three, four, or any number. For example, I television cameras or photosensor arrays might form a circle or semicircle around a region that contains objects of interest, or a two-dimensional array of I cameras might be used.

6.1 Structures to process *I* identical input images

The above systems for two input images can be extended to handle I input images, all the same size, simply by multiplying the number of wires to the pyramid or array that handles the higher levels by the number of lower level truncated pyramids that must be handled. Extra fan-in hardware is then added as needed.

Alternatively, the base array of the combining pyramid can be made suitably larger, to minimize or eliminate extra fan-in hardware. For example, link each of four subsets of sixteen truncated pyramids to a different computer in a 4×4 subarray of the combining pyramid's base. All sixteen will then be combined at the next layer of the pyramid, possibly with intermediate processing along the way.

6.2 Structures to process *I* images of different sizes

For certain problems it might be desirable to have several inputs of different sizes. For example, one or several large arrays can be used to monitor a wide angle of vision, along with several smaller arrays concentrated at the centre.

The constructions just described might be used, adding fan-in hardware as needed. In addition, the system might link different sized pyramids at different layers. For example, a 1024×1024-based pyramid and a 128×128-based pyramid might be linked 3 layers up from their bases, where the first was 128×128 and the second was 16×16. The larger might be linked into the 32×32 base of a combining pyramid, the smaller into the 4×4 interior layer.

7 PYRAMIDS THAT HANDLE SEQUENCES OF IMAGES, AS FOR MOVING PICTURES

Moving pictures are captured with a sequence of image frames, as by a television or ciné camera. Each image is the same size; all images overlap one another completely. (This ignores pans and zooms, but they can be handled by the region-combining techniques described above.) Such a sequence can be handled in any of the following ways.

All images are input to a single pyramid's base, each processed in pipeline fashion moving up the pyramid. Information can be retained from an image and combined with information from subsequent images (it is up to the programmer how this is handled). To do this efficiently, the set of processes must be load-balanced across layers, so that every layer's sequence of processes takes almost exactly the same amount of time. Also, virtually all processing must be bottom-up, since the pipelining cannot spare much time from the continuing stream of images to interrupt the flow and to move down the pyramid.

These are difficult to achieve within the extremely stringent real-time constraints. Therefore it might be preferable to augment the pyramid.

First, each computer might be made a multiprocessor. For example, four or eight processors might share a common memory. Now each processor could be responsible for a different one of four or eight frames (the oldest frame being abandoned when a new frame is to be processed). Alternatively, one processor might be responsible for combining old frames into "recent-history" images, while the other processors were working on the last few images. This effectively allows the system to work on each image for longer than 30 ms (e.g. using eight processors means that each processor could spend up to 240 ms on the image assigned to it).

Secondly, the base array might be linked to several (e.g. four or eight) identical pyramids. Now each pyramid could work on a different frame. This would not have the flexibility of a single pyramid with multiprocessors, since information being uncovered in one pyramid would not be accessible to another. It would therefore be necessary to link all the separate pyramids to a combining pyramid or network—as described for binocular and I-ocular pyramids.

8 MULTIMODAL PYRAMIDS

A pyramid is, potentially, an attractive structure for processing and perceiving images input from several different sense modalities. For example, the system might input images from two eye-like sensors (television cameras), two ear-like sensors (acoustic filters) and several finger-like touch sensors (arrays of pressure gauges).

In all these cases, information must first be input with enough resolution so that the details needed for perception are not muddied and lost. Thus the

Multiple-image and multimodal pyramids 141

eye-like sensors should input image arrays that are at least 128×128, and very possibly 1024×1024 or even larger. The ear-like sensors should input either (a) vectors (that is, one-dimensional arrays) at a rate of roughly 10 000 to 50 000 samplings per second, or 5000 to 20 000 per word, or (b) two-dimensional arrays (time vs. frequency) with roughly 40×40 per word. The finger-like sensors will typically input much smaller arrays, probably 10×10 or 20×20 at most, and often only 1×1.

Extensions to the constructions for binocular and *I*-ocular pyramids given above appear to be indicated as attractive ways to handle such multimodal images. In general, the transformed image information should be merged together in pairs, each pair at the appropriate level. But two new problems arise.

The one-dimensional transformed speech image must be combined with two-dimensional transformed images.

The ways that images map onto the external world will differ markedly, in ways that must be taken into account when deciding how to combine images. For example, the images from touch may be oriented in the third dimension (that is, from the finger touching a plane that is not in the plane projected to the eyes). Also, the acoustical images will typically refer to the entire physical world in the sense that they are heard from all directions—although occasionally they will be localized, as when the sound of a robin helps identify a particular seen bird as a robin.

8.1 Combining two-dimensional arrays for vision and touch

When the several arrays are all two-dimensional the procedures described above for *I*-ocular images of different sizes and resolutions appear to be indicated for multimodal images as well.

The major difference lies in the fact that radically different kinds of procedures will be needed to handle the different kinds of features and combinations of features that must be assessed in the different sense modalities. Since these pyramids use general-purpose processors, the hardware, if well designed, should be adequate; the problems are the standard ones of developing the appropriate algorithms and programs.

Possibly the most difficult problem lies in deciding which pieces of information from each of several arrays should be brought together and combined. This too is largely a programming problem. But it may well be that information is dispersed in different ways in the different arrays. For example, a touch array might get information about the general smoothness or flatness of a surface, and this information might well be of use almost everywhere in the array of visual information with which it is combined.

This suggests that additional hardware might effectively be used to link the lower-level arrays whose transformed images are to be combined with the higher-level combining array. For example, a fan-out network might directly link a small array containing information derived from touch sensors with all the computers in a larger array, one that is large enough to receive information derived from visual images. Alternatively, a procedure can be coded to broadcast this information via the links that join the computers in the higher-level network.

8.2 Combining one-dimensional with two-dimensional images

There appear to be several possibilities for combining two-dimensional with one-dimensional images, in particular sound (e.g. speech; animal calls and songs; noises like wind, rain, explosions, motors).

Probably the simplest approach is the following: The one-dimensional array (e.g. a vector of acoustical samples) is input to the base of a two-dimensional pyramid, and successively transformed while moving up through this structure. At whatever layer is deemed most appropriate, this information is transferred into a conventional computer into which is also transferred the transformed image from a three-dimensional pyramid that has been processing a two-dimensional image (e.g. a television picture).

Alternatively, an MIMD network might be used instead of the single conventional computer. Information that should be broadcast (e.g. the sound of rain) can be input to whatever single computer the subarray in which it was recognized is linked, and then passed on to the other computers over the links internal to the network, or a fan-out network could broadcast directly to all.

8.3 Combining rectangular-framed (including one-dimensional) images

It might be well to consider using rectangular pyramids for vision, with base arrays of the sort used in Cinemascope and, in less pronounced form, in most television and photographic images. The rectangle appears to fit the meaningful portion of an image (e.g. an expanse of people, animals or buildings extending along the Earth's surface, or—when the rectangle is rotated 90°—a person, or a tall building or tree). Binocular images will often naturally be rectangular, and a rectangle more nearly approximates the image input by the human being's two eyes than does a square.

Multiple-image and multimodal pyramids 143

A Cinemascope-like rectangular image will converge at some layer to a one-dimensional array. This might be the appropriate layer for merging information together (considered purely from the hardware point of view— it might be quite inappropriate from the more important perspective of the information that has been abstracted that the system must combine).

9 TOWARD A GENERAL-PURPOSE MULTIPYRAMID

A variety of different constructions have been proposed. Each has been relatively specific to the problem posed: binocular or I-ocular vision, a stream of images of objects in motion, multimodal perception. What would be the best design for a multicomputer that could handle any and all of these problems with reasonable efficiency?

One attractive alternative would use several augmented pyramids, each with B-bit computers that are reconfigurable to several computers of the same smaller size, or even of different sizes, or to B one-bit computers. These could then be assigned, as appropriate, to the different input images.

A rectangular shape (probably either $N \times 2N$ or $N \times 4N$) might well be preferable. Thus the pyramid's base might best be 256×512 or 256×1024.

This reconfigurable multipyramid should then be augmented by linking it at one of its higher levels to an MIMD network of appropriate size and shape—probably a small pyramid, a Petersen graph, a Singleton graph or a reasonably dense compound like the one formed from 51 Singleton graphs, or an augmented tree like a hypertree or a de Bruijn network.

10 SUMMARY

This chapter has examined how hardware pyramids might be designed to handle problems where several images are input, either simultaneously or in sequence.

The simplest situation is one where two images are input that are of identical size from the identical region. This is the case for binocular vision with complete overlap. A variety of combinations of pyramids and truncated pyramids are proposed to handle such situations, where one pyramid is linked at a middle-level layer into the second, or two pyramids are linked into a third combining pyramid, into a conventional computer or into an MIMD network of more powerful computers.

When the two images do not overlap completely, or do not overlap at all, the combining pyramid must have an appropriate rectangular shape into which the two abstracted images from the lower-level pyramids can be merged in such a way as to preserve their relative locations.

I-ocular inputs, of more than two images of either identical or different sizes, can be handled in essentially the same ways. Different layers of the combining pyramid can be used, as appropriate, to converge different sized abstracted images from the different pyramids in order to combine them.

Sequences of images, all of an identical format (as are the successive frames from a television camera) can be handled in a single pyramid. But because of severe time constraints, either a pyramid whose individual computers have several processors, or an input array that is linked to several separate pyramids, which in turn are linked to a combining pyramid or network, appears to be preferable.

Multimodal inputs can be handled with hardware structures similar to those described for I-ocular images of different sizes. Different kinds of procedures will be needed to process information in each sense modality, but—so long as well-designed general-purpose computers are used—this reduces to a programming problem. Multimodal inputs also pose new problems, when one of the modes is one-dimensional (as is one of the standard encodings for speech), and also when locality of information differs greatly from image array to image array. For example, spoken speech or the sound of rain is essentially without location; the sound of a bird should be broadcast everywhere to help identify the appropriate bird at a specific location. Additional links and fan-out broadcast networks might be indicated in such cases. Rectangular arrays for images other than the one-dimensional (therefore rectangular) sound array also appear to be desirable, to give a more natural framing and a more natural merging.

This paper has examined a variety of augmented pyramid, multiple-pyramid, and multiple-pyramid-plus-network structures that might be used to handle the different types of problems where more than one image, whether of the same or different types or modalities, must be processed. There appears to be enough similarity among these different problems so that a relatively general-purpose multipyramid might be designed to handle all of them with reasonable efficiency. Among the most attractive would appear to be a rectangular pyramid of reconfigurable B-bit computers ($B=16$ or $B=32$) linked at a higher layer to one of the relatively dense MIMD networks described above.

ACKNOWLEDGMENT

This research was partially supported by the National Science Foundation under grant DCR-8302397.

REFERENCES

[1] Dyer, C. R. (1982). Pyramid algorithms and machines. In *Multicomputers and Image Processing* (ed. K. Preston and L. Uhr), pp. 409–420. Academic Press, New York.
[2] Hanson, A. R. and Riseman, E. M. (1974). Pre-processing cones: a computational structure for scene analysis. *COINS Tech. Rep.* 74C-7, Univ. Massachusetts.
[3] Hanson, A. R. and Riseman, E. M. (1978). Visions: a computer system for interpreting scenes. In *Computer Vision Systems* (ed. A. R. Hanson and E. M. Riseman), pp. 303–333. Academic Press, New York.
[4] Rosenfeld, A. (ed.) (1984). *Multi-Resolution Image Processing and Analysis*. Springer, Berlin.
[5] Tanimoto, S. L. (1976). Pictorial feature distortion in a pyramid. *Comp. Graph. Image Process.* 5, 333–352.
[6] Tanimoto, S. L. (1983). A pyramidal approach to parallel processing. In *Proc. 10th Ann. Int. Symp. on Computer Architecture, Stockholm*, pp. 372–378.
[7] Tanimoto, S. L. (1984). A hierarchical cellular logic for pyramid computers. *J. Parallel and Distributed Comp.* 1, 105–132.
[8] Tanimoto, S. L. and Klinger, A. (eds) (1980). *Structured Computer Vision* Academic Press, New York.
[9] Uhr, L. (1972). Layered "recognition cone" networks that preprocess, classify and describe. *IEEE Trans. Comp.* 21, 758–768.
[10] Uhr, L. (1978). "Recognition cones" and some test results. In *Computer Vision Systems* (ed. A. Hanson and E. Riseman), pp. 363–377. Academic Press, New York.
[11] Uhr, L. (1984). *Algorithm-Structured Computer Arrays and Networks: Architectures and Processes for Images, Percepts, Models, Information*. Academic Press, New York.
[12] Uhr, L. (1980). Psychological motivation and underlying concepts. In *Structured Computer Vision* (ed. S. Tanimoto and A. Klinger), pp. 1–30. Academic Press, New York.
[13] Unger, S.H. (1958). A computer oriented toward spatial problems. *Proc. IRE*. 46, 1744–1750.
[14] McCormick, B.H. (1963). The Illinois pattern recognition computer ILLIAC III. *IEEE Trans. Comp.* 12, 791–813.
[15] Slotnick, D.L., Borck, W.C. and McReynolds, R.C. (1962). The Solomon Computer. *Proc. AFIPS FJCC* pp. 87–107.
[16] Duff, M. J. B. (1976). CLIP4: a large scale integrated circuit array parallel processor. In *Proc. 3rd Int. Joint Conf. on Pattern Recognition*, pp. 728–733.
[17] Reddaway, S.F. (1978). DAP—a flexible number cruncher. In *Proc. 1978 LASL Workshop on Vector and Parallel Processors, Los Alamos*, 233–234.
[18] Batcher, K. E. (1980). Design of a massively parallel processor. *IEEE Trans. Comp.* 29, 836–840.
[19] Flynn, M. (1972). Some computer organizations and their effectiveness. *IEEE Trans. Comp.* 21, 948–960.
[20] Tanimoto, S. L. (1981). Towards hierarchical cellular logic: design considerations for pyramid machines. *Comp. Sci. Dept Tech. Rep.* 81-02-01, Univ. Washington.

[21] Uhr, L. (1981). Converging pyramids of arrays. In *Proc. Workshop on Computer Architecture for Pattern Analysis and Image Data Base Management*, pp. 31–34. IEEE Computer Society Press.
[22] Cantoni, V., Ferretti, M., Levialdi, S. and Maloberti, F. (1985). A pyramid project using integrated technology. In *Integrated Technology for Parallel Image Processing* (ed. S. Levialdi), pp. 121–132. Academic Press, London.
[23] Schaefer, D. H. (1985). A pyramid of MPP processing elements—experiences and plans. In *Proc. 18th Int. Conf. on System Sciences, Honolulu*.
[24] Uhr, L. (1983). Pyramid multi-computer structures, and augmented pyramids. In *Computing Structures for Image Processing* (ed. M. J. B. Duff), pp. 95–112. Academic Press, London.
[25] Uhr L., *Perception and Multi-Computers*. In preparation.
[26] Dyer, C. R. (1981). A quadtree machine for parallel image processing. *Information Engng. Dept. Tech. Rep.* KSL 51, *Univ. Illinois, Chicago Circle*.
[27] Sandon, P. A. (1985). A pyramid implementation using a reconfigurable array of processors. In *Proc. Workshop on Computer Architecture for Pattern Analysis and Image Data-base Management*. IEEE Computer Society Press.
[28] Siegel, H. J. (1981). PASM: a reconfigurable multimicrocomputer system for image processing. In *Languages and Architectures for Image Processing* (ed. M. J. B. Duff and S. Levialdi), pp. 257–265. Academic Press, London.
[29] Siegel, H. J., Siegel, L. J., Kemmerer, F. C., Mueller, P. T., Smalley, H. E. and Smith, S. D. (1981). PASM: a partitionable SIMD/MIMD system for image processing and pattern recognition. *IEEE Trans. Comp.* **30**, 934–947.
[30] Pfeiffer, J. J. (1985). Integrating low level and high level computer vision. In *Proc. Workshop on Computer Architecture for Pattern Analysis and Image Data-base Management*. IEEE Computer Society Press.
[31] Tanimoto, S. L. (1985). An approach to the iconic/symbolic interface. In *Integrated Technology for Parallel Image Processing* (ed. S. Levialdi), pp. 31–38. Academic Press, London.
[32] Ahuja, N. and Swami, S. (1984). Multiprocessor pyramid architectures for bottom-up image analysis. In *Multiresolution Image Processing and Analysis* (ed. A. Rosenfeld), pp. 38–59. Springer, Berlin.
[33] Sharma, M., Patel, J. H. and Ahuja, N. (1985). NETRA: an architecture for a large scale multiprocessor vision system. In *Proc. Workshop on Computer Architecture for Pattern Analysis and Image Data-base Management*. IEEE Computer Society Press.
[34] Uhr, L. (1983). Augmenting pyramids and arrays by compounding them with networks. In *Proc. Workshop on Computer Architecture for Pattern Analysis and Image Data Base Management*, pp. 162–169. IEEE Computer Society Press.
[35] Uhr, L. (1985). Augmenting pyramids and arrays by embossing them into optimal graphs to build multicomputer networks. In *Integrated Technology for Parallel Image Processing* (ed. S. Levialdi), pp. 19–30. Academic Press, London.
[36] Uhr, L. (1985). Pyramid multi-computers, and extensions and augmentations. In *Algorithmically Specialized Computer Organizations* (ed. D. Gannon, H. J. Siegel, L. Siegel and L. Snyder). Academic Press, New York.
[37] Bermond, J.-C., Delorme, C. and Quisquater, J.-J. (1983). Strategies for interconnection networks: some methods from graph theory. *Manuscript M58, Philips Res. Lab., Brussels*.

[38] Hoffman, A. J. and Singleton, R. R. (1960). On Moore graphs with diameter 2 and 3. *IBM J. Res. Dev.* **4,** 497–504.
[39] Goodman, J. R. and Sequin, C. H. (1981). Hypertree, a multiprocessor interconnection topology. *Comp. Sci. Dept. Tech. Rep.* 427, *Univ. Wisconsin.*
[40] de Bruijn, D. G. (1946). A combinatorial problem. *Kon. Ned. Acad. Wetenschappen Amsterdam, Proc. Sect. Sci.* **49, 7,** 758–764.

Chapter Nine
Algorithm-Guided Design Considerations for Meshes and Pyramids

Q. F. Stout

1 INTRODUCTION

This paper uses an analysis of algorithms to suggest some desirable features of machines to be used for image understanding. The emphasis is on modifying mesh and pyramid computers, defined in the next section, because such machines are currently being used for a variety of image-processing tasks. Meshes and pyramids are well suited to several image-processing problems, but it will be shown they seem to be less suitable for higher-level tasks of image understanding.

Ideally, one designs a machine for image understanding by starting with a representative collection of good image-understanding algorithms, and then developing an architecture that efficiently runs the algorithms. Unfortunately, such algorithms do not yet exist, so we instead examine algorithms that have features that seem to be related to features in future image-understanding algorithms. Traditional image problems such as filtering and the Hough transform are considered, as well as problems from graph theory and geometry. Graph theory is considered because standard artificial intelligence constructs such as semantic networks, AND/OR trees, constraint propagation, neural nets, etc. are all based on graph theory [1].

Geometry is considered because the initial stages of image processing deal with the "iconic" pixel input, while the final stages of image understanding will probably deal with a "symbolic" representation. The symbolic representation may include an abstract representation of objects (e.g. edges and regions) and properties of the objects (e.g. their size and relative orientation). Regardless of whether or not geometry is an appropriate link between iconic and symbolic representations, the computational requirements of determining geometric properties of objects in the image should be representative of some of the computational requirements of determining other properties of the image.

Some of the features discussed here are present on some image-processing machines, but none are universally present. Because the emphasis is on future machines, no attempt has been made to include a comprehensive survey of which current machines have which features.

2 DEFINITIONS

For our analyses the input will alway be an $n \times n$ array of pixels, for some integer n. This will be called an *image of size* n^2. Currently $n = 512$ is common, with increasing interest in n as large as 4000.

A *mesh of size* n^2 has n^2 *processing elements* (*PEs*) arranged in a square lattice. We use the l_1 of "four nearest neighbours" interconnection scheme in which PEs at (i_1, j_1) and (i_2, j_2) are connected via unit-time communication links if and only if $|i_1 - i_2| + |j_1 - j_2| = 1$ (see Fig. 1a). Other interconnection schemes are also used, such as l_∞ or "eight nearest neighbours" in which (i_1, j_1) and (i_2, j_2) are connected if and only if max $\{|i_1 - i_2|, |j_1 - j_2|\} = 1$, or "six nearest neighbours" based on a hexagonal tessellation of the plane.

A *pyramid of size* n^2, where n is a power of 2, actually has $\frac{4}{3}n^2 - \frac{1}{3}$ PEs arranged in $1 + \lg(n)$ layers. (lg is $\log_2$.) The base is a mesh of size n^2, above the next layer is a mesh of size $\frac{1}{4}n^2$, then a mesh of size $\frac{1}{16}n^2$, and so on to the

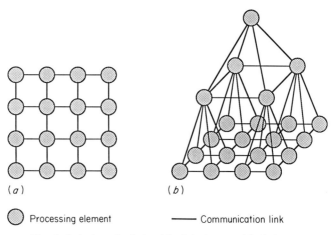

○ Processing element   —— Communication link

Fig. 1. (a) A mesh of size 16. (b) A pyramid of size 16.

Algorithms for meshes and pyramids 151

apex of size 1. Each PE is connected to its mesh neighbours on the same level, four children on the level below (except for PEs on the base), and its parent on the level above (except for the apex) (see Fig. 1*b*). If all the mesh connections are omitted, then the resulting machine is a *tree of size* n^2. A tree is a poor image processing architecture, but it will be briefly used for comparison purposes.

Throughout the following, image, mesh or pyramid without an explicit size mentioned will mean the appropriate object of size n^2.

We concentrate on analysing running time of a task, ignoring the input/output time since the intent is that the entire analysis be performed on the parallel machine. Sometimes efficiency (PE utilization) or speed-up (ratio of time on parallel machine to time on a serial one) are analysed, but here running time seems the most appropriate. In particular, adding PEs to image-processing machines may incur sublinear costs, in which case the standard efficiency and speed-up measures do not correctly measure cost performance.

Let $T(N)$ and $G(N)$ be arbitrary positive functions. Then $T = O(G)$ if there are constants $d, m > 0$ such that $T(n) \leq d \star G(n)$ whenever $n \geq m$, $T = \Omega(G)$ if there are constants $c, m > 0$ such that $c \star G(n) \leq T(n)$ whenever $n \geq m$, $T = \Theta(G)$ if $T = O(G)$ and $T = \Omega(G)$, and $T = o(G)$ if for any constant $d > 0$ there is a constant $m > 0$ such that $T(n) \leq d \star G(n)$ whenever $n \geq m$. For example, $2n \star \sin^2(n) + 17 = O(n)$, $2n \star \sin^2(n) + 17 = \Omega(1)$, $3n^2 - 2n2 = \Theta(n^2)$, and $\log_2 (n) = o(n^\varepsilon)$ for any $\varepsilon > 0$. $T = O(G)$, $T = \Omega(G)$, $T = \Theta(G)$ and $T = o(G)$ are sometimes read as "*T* is of order no greater than *G*", "*T* is of order no less than *G*", "*T* is of order *G*" and "*T* is of order strictly less than *G*" respectively.

Since $\log_a (n) = \log_a (b) \star \log_b (n)$, $\log_a (n) = \Theta(\log_b (n))$ whenever $a, b > 1$. Because of this, the base of a logarithm is usually omitted inside of an O, Ω, Θ or o. *T* is *poly-log* if $T = O(\log^k)$ for some k.

3 SIMD/MIMD

Most current meshes and pyramids have bit-serial PEs used in SIMD mode, meaning that all PEs execute the same instruction (Single Instruction), but apply it to their own data (Multiple Data). An alternative is MIMD, whereby different PEs may execute different instructions (Multiple Instruction). The fact that most parallel image-processing machines are SIMD distinguishes them from general parallel machines. In fact, in a recent survey of parallel-processing projects, not one project emphasized work on SIMD machines [2]. While this survey was obviously incomplete, and there are SIMD projects being pursued, it does illustrate that the preponderance of work on parallel machines and algorithms is aimed at MIMD machines.

An interesting feature of real SIMD machines is that they are ignored by the standard theoretical models of parallelism. For example, the early models of meshes and pyramids assumed that the PEs were copies of a finite state automaton F, where F was independent of n [3–7]. Such PEs cannot, in general have enough memory capacity to store n or their coordinates, and there is no straightforward way for them to, say, repeat a loop n times. Real SIMD machines have a controller that can count loop iterations. Collections of finite state automata can simulate this ability, but it involves considerable complication [6].

More recent theoretical models replace F with a processor capable of storing its coordinates and doing calculations on words of length $\Omega(\log(n))$ in unit time [8–12]. This model is usually an MIMD machine, but is almost invariably used only in SCMD (Single-Code, Multiple-Data) mode. In SCMD mode the PEs run the same program at about the same rate, but because of data differences, different PEs may be executing different branches. Usually SCMD mode involves synchronization through some form of signal or message passing.

Any SCMD program for a machine can be converted to one for an SIMD machine with masking ability and the same interconnection scheme. The conversion can be accomplished by first numbering the lines of the original program, and setting a variable "program__counter" in each PE to be 1°. (The superscript o is for "odd".) The controller repeatedly iterates odd and even "cycles" until the problem is finished, where each odd cycle is of the form:

> **if** program__counter $= 1^o$ **then** simulate line 1;
> **if** program__counter $= 2^o$ **then** simulate line 2;

where line 1 is simulated by

> execute line 1;
> program__counter $\leftarrow 2^e$;

if line 1 is not a jump, and by

> program__counter $\leftarrow j^e$; (where j is the jump destination line)

otherwise. Even cycles are the same except that ^{e} and ^{o} are interchanged.

The number of cycles needed is the number of instructions executed in the SCMD version, and the time for a single cycle is independent of n. Hence the SIMD and SCMD versions have the "same" time in an O-notational analysis, despite the fact that the conversion is so poor as to be useless.

Algorithms for meshes and pyramids

While the above conversion is not used, it does illustrate the importance of conditional (also called masked) execution in SIMD mode, in which a PE either ignores the instruction if the condition is false, or else performs it and discards the result. Some simple meshes and systolic arrays do not have conditional execution, but almost all nontrivial algorithms require it since it is conditional execution that most distinguishes a computer from a calculator.

While most SIMD meshes and pyramids have conditional execution, extensive use of it causes inefficiency. For example, suppose each pixel has been classified into one of four types, and a case structure is encountered as in Fig. 2, where the code to be executed depends upon the pixel's classification. If blocks A, B, C and D have the same execution time, then each PE executes only $\frac{1}{4}$ of the issued instructions.

In some cases, this can be ameliorated by a proper instruction set. For example, in many divide-and-conquer algorithms there are parts depending on the PE's position in which the only difference between branches are the neighbour the PE should communicate with. For example, it may be that if the PE's first coordinate is less than $\frac{1}{2}n$ it will send data to its east neighbour, while otherwise it sends it to its west neighbour. In this situation switch-selectable I/O ports can be used in which switch settings are computed and the code is combined and rewritten in terms of communication with the selected neighbour.

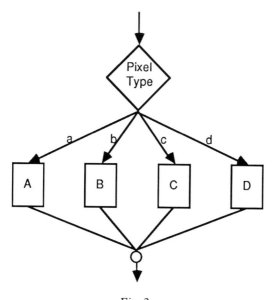

Fig. 2.

In more general algorithms, it is harder to eliminate this inefficiency of the SIMD framework. If the code in blocks A, B, C and D of Fig. 2 is to be executed repeatedly, then one possibility is to load the microcode into each PE, with conditional jumps to execute the appropriate block. This increases the storage needed per PE, and synchronization is needed at the end of the blocks. The execution time of this part will be $\frac{1}{4}$ of that before, although there is the additional time to load the microcode. The machine is just barely an MIMD machine. It still has fairly simple PEs, only one instruction decoder, and the bulk of the program is still stored in only one location, as opposed to the multiple decoders and program copies in MIMD machines.

In pyramids the inefficiency of different code blocks is compounded by the fact that a PE's function may depend on its level, as well as the data beneath it. Some early pyramid-like models did not require that all levels be performing the same operations [13,14] and it has been suggested that pyramids could return to such models, perhaps with each level operating in SIMD mode with a different controller on each level.

Possibilities such as these are compromises between SIMD hardware efficiency and MIMD flexibility and software efficiency (PEs do not wait for long periods while irrelevant instructions are being broadcast to them). Insightful compromises will be important for a long time, but it seems that the trend is definitely towards MIMD machines, at least in SCMD mode. Artificial intelligence researchers will push for MIMD image-understanding machines because they will have MIMD algorithms, and because they will not tolerate the inefficiency and obnoxiousness of translating MIMD algorithms to SIMD machines.

4 INTERPROCESSOR COMMUNICATION

This section is concerned with analysing the communication between PEs required to solve various classes of problems. This is a parameter that is used extensively in analyses of VLSI and parallel computers. For example, given a problem in which information from an arbitrary pair of PEs must be combined, then any algorithm to solve the problem must have a worst case time at least half of the communication diameter of the machine. The *communication diameter* of a machine is the smallest number D such that any pair of PEs is joined by a path of D or fewer communication links. (We assume that all links are bidirectional and take unit time.) A mesh of size n^2 has a communication diameter of $2n-2$, and a pyramid of size n^2 has a communication diameter of $2 \lg(n)$.

The problems to be considered fall into four classes, called mesh-local, pyramid-local, perimeter-bound and strongly global. These classes are not

Algorithms for meshes and pyramids 155

exhaustive, but seem to capture the communications requirements of most image related problems. We define each class, analyse how meshes and pyramids perform on them, and discuss some possible improvements. Analyses will concentrate on the communication aspects of the problem, which for most of the problems determines the running time.

For all the problems considered below, the data are initially in place, all PEs start simultaneously, and, if needed, PEs initially contain their coordinates and n. The input is an image of size n^2 stored naturally one pixel per PE in a mesh of size n^2, or one pixel per base PE in a pyramid of size n^2. The result is either an $n \times n$ matrix, as would occur in image rotation, or some collection of values to be computed by some specified PE, as would occur if the apex of a pyramid should end up containing the number of fish, and the number of bicycles, in the image. All analyses are worst case.

4.1 Problem classes

Mesh-local problems are ones in which there is a distance D, independent of n, such that the final values at any pixel p depend only upon the initial values at those pixels q such that $d(p, q) \le D$, where distance is measured by the Euclidean metric. (For distances, pixels are identified with their integer coordinates.) An example is median filtering, in which each pixel value is to be replaced with the median of its initial value and the initial values of its eight nearest neighbours. (There must be a suitable definition for filtering of pixels along the edges.) Other mesh-local problems include average filtering and various edge and corner detection problems.

Pyramid-local problems are problems that involve combining information from PEs arbitrarily far apart, and for which, given any two disjoint squares S and T of s^2 pixels, the PEs working on S and T need to exchange $O(s \star \log(s)/D(S,T))$ words of information, where $D(S,T)$ is the smallest Euclidean distance between a pixel in S and a pixel in T. Pyramid-local problems include determining the average grey level of the image, labelling connected components where the only components are circles or squares (or certain other figures), determining the minimum distance between any pair of pixels in some marked set, deciding if the black pixels are a convex set, and finding the median. All of these problems can be solved in poly-log time on a pyramid [15–19], which justifies the name. It is easy to show that the distance or labelling problems cannot be solved in poly-log time on a tree because too much data must leave each square.

Perimeter-bound problems are ones in which, given any square of pixels, the PEs at those pixels need only exchange an amount of information proportional to the perimeter of the square with the outside world, but much of this information has to travel far. More formally, a perimeter-bound problem is one in which, for any square of s^2 pixels, no more than $O(s)$ words of information need to be exchanged between the PEs corresponding to the square and the PEs corresponding to the rest of the input, and for which there is a constant $\varepsilon > 0$ such that, for any $s \le \frac{1}{2}n$, there are two squares of s^2 pixels touching opposite edges of the image for which the corresponding PEs must exchange $\Omega(s^\varepsilon)$ words of information. The second requirement guarantees that perimeter-bound problems are not pyramid-local.

Perimeter-bound problems include general component labelling, deciding the convexity of each figure, distances to nearest neighbours, and several graph problems that have an adjacency or weight matrix as input. To show that component labelling satisfies the first constraint on perimeter-bound problems we use the standard divide-and-conquer solution [20]. The goal of component labelling is to assign a label to each black pixel, with two pixels receiving the same label if and only if there is a connected path of black pixels between them. First components are labelled in each quadrant, ignoring the other quadrants. The only components that are mislabelled (have two different labels for the same component) are those in two or more quadrants, (see Fig. 3a). However, such components must lie on the border of the quadrants, so labels along the borders are combined and recalculated. Any changes are then sent back to the quadrants to redo the changed labels. This algorithm shows that only $O(s)$ information need flow between a square of size s^2 and the outside world. Further, Fig. 3(b) shows that the second condition, with $\varepsilon = 1$, is also satisfied. The Ys on the rows above and below an X will have the same label if and only if the X is black, so information concerning the blackness of the Xs must be transmitted to the Ys.

Strongly global problems are such that if the image is divided horizontally (or vertically) through its middle, then the two halves must exchange $\Omega(n^2)$ words of information. Strongly global problems include image rotation,

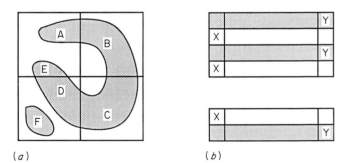

(a) (b)

Fig. 3. (a) Component labels as assigned in quadrants. (b) Ys are black, Xs are black or white.

Algorithms for meshes and pyramids 157

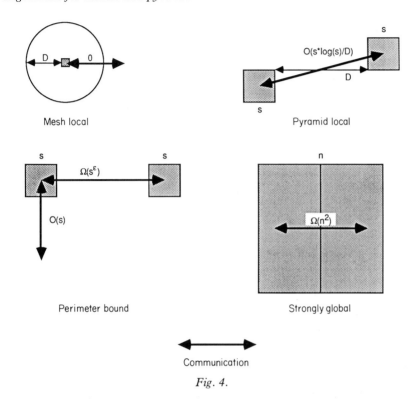

Fig. 4.

convolution, many definitions of the Hough transform, sorting, matrix multiplication and most graph problems if the input is unsorted edges. This class spans the widest variation in computation requirements.

Figure 4 illustrates the communication characteristic of these classes.

4.2 Mesh performance

Obviously a mesh of size n^2 can complete the communications for a mesh-local problem of size n^2 in $\Theta(1)$ time. For all the other classes, the fact that communication is required between PEs arbitrarily far apart ensures that the mesh must take $\Omega(n)$ time. Each of the pyramid-local and perimeter-bound problems mentioned can be solved in $\Theta(n)$ time on a mesh [9,11,21].

One can concoct pyramid-local and perimeter-bound problems that cannot be solved in $\Theta(n)$ time on a mesh, but these problems do not seem to occur naturally. One way such a problem may come about naturally is if one asks an NP-hard question about the arrangement of pieces of the image, for example an optimal matching between the large connected components of two images. In this case the reduction of an image to some components may yield a significant data reduction, but the remaining problem is so hard that it still takes a long time. Graph-optimization problems will take a long time no matter what computer they are on.

Finally, for a mesh to solve a strongly global problem, there are only n wires between the two halves, so it takes $\Omega(n)$ time to move $\Omega(n^2)$ words. Image rotation, sorting, and several edge-input graph algorithms can be done in $\Theta(n)$ time [12,18], but some problems cannot be solved this fast. For example, the Hough transform does one calculation for each pair of edge pixels. There can be $\Theta(n^4)$ pairs, so it takes $\Omega(n^2)$ time on a mesh of size n^2 (this lower bound is attainable).

4.3 Improved meshes

From the above comments, we see that meshes are essentially optimal for the simple mesh-local problems, while if they are to solve strongly global problems in $o(n)$ time then drastic changes are needed. Simple changes, such as adding a global bus or even a bus for every row and column, do not add enough bandwidth to move data from one half to the other. Further, for computationally intensive problems such as the Hough transform, the ratio of computations to PEs shows that to solve the problem in $o(n)$ time the number of PEs must grow faster than the input size.

The pyramid-local and perimeter-bound problems have the most potential for speed-up on modified meshes. A global bus can reduce the time for finding the average grey level from $O(n)$ to $O(n^{2/3})$ on a mesh of size n^2, and a few other pyramid local problems can be similarly improved by a global bus [22,23]. Global bus algorithms tend to do mesh processing on disjoint squares of $n^{4/3}$ PEs, taking $\Omega(n^{2/3})$ time, interspersed with periods of global broadcasting. Since there are $\Theta(n^{2/3})$ squares, if each square broadcasts at least once then the total broadcast time is also $\Omega(n^{2/3})$.

For perimeter-bound problems a global bus is less successful. If mesh processing is done on squares of size m^2 then the mesh processing will take $\Omega(m)$ time. Each square may have to broadcast $\Theta(m)$ values, and there are $\Theta((n/m)^2)$ squares, so the broadcast time is $\Omega(n^2/m)$. This is minimized when $m = n$; that is, when the standard mesh is used with no broadcasting. A bus per row and column can provide greater broadcasting capacity, reducing the time of some perimeter-bound problems to $o(n)$ [24]. Other bussing strategies can offer slightly greater improvement for comparable construction costs [25]. Similar improvements can be obtained by adding a few wires, for example one might connect the PE at $(i \star n^{1/2}, j \star n^{1/2})$ to the PEs at

Algorithms for meshes and pyramids

$((i \pm 1){\star}n^{1/2}, j{\star}n^{1/2})$ and $(i{\star}n^{1/2}, (j \pm 1){\star}n^{1/2})$, using these new wires much like busses.

However, all of these improvements carry a higher construction cost, and demand substantially more complex algorithms to make full use of their capabilities. The algorithms are sufficiently complicated that they would almost certainly be run only on an MIMD computer, and it is probable that in many cases a programmer would ignore an extra feature, rather than take the time needed to determine how to fully use it in a particular application.

Another way to obtain faster mesh solutions to pyramid-local and perimeter-bound problems is to note that it is the communication diameter of the mesh that forces $\Omega(n)$ time. If a smaller mesh is used, say of size $m \times m$, then the communication diameter is $\Theta(m)$. To compensate for fewer PEs, each PE should have more memory, storing a $(n/n) \times (n/m)$ subsquare of the image. In [10] it is shown that using $m = n^{2/3}$ allows all the pyramid-local and perimeter-bound problems mentioned to be solved in $\Theta(n^{2/3})$ time. Note that $m = n^{2/3}$ is the point at which the communication diameter equals the time a PE needs to examine its own data. However, it should be noted that this machine takes $\Omega(n^{2/3})$ time on mesh-local problems, and for strongly global problems it takes $\Omega(n^{4/3})$ time to move $\Omega(n^2)$ words between halves.

Such a machine would almost certainly be MIMD. This machine also requires more complex algorithms, since they must blend local serial and global parallel processing, but the programmer is rewarded by having PEs with far more memory and probably a more powerful instruction set. Since the machine has fewer PEs and wires than a standard mesh, the cost of such a machine should compare favourably with a standard mesh.

4.4 Pyramid performance

The pyramid computer is similar to the mesh in its behaviour on mesh-local problems, and it solves all the mentioned pyramid local problems in poly-log time. In fact, except for the median, all of the mentioned pyramid local problems can be solved in $\Theta(\log(n))$ time [17,19], and it is an open question to determine how fast a pyramid can find the median.

It may seem that a pyramid can quickly solve some strongly global problems since each half of the base has $\Theta(n^2)$ wires to the outside. However, most of these are to the level above. If a pyramid is sliced straight down just to one side of the apex, then only $2n$ wires are cut. This is only twice as many as are cut on the base, so again $\Omega(n)$ time is needed to move $\Omega(n^2)$ words.

Perimeter-bound problems pose the most interesting lower bounds. If a perimeter-bound problem requires moving $\Omega(n^\varepsilon)$ words from a square of $(\frac{1}{2}n)^2$ pixels along one edge to PEs on the opposite edge, then the pyramid needs $\Omega(n^{\varepsilon/2})$ time [26]. This bound arises because data must travel up and down the pyramid to move rapidly from one side to another, but if much data are trying to make this transit then the apex is a bottleneck. One can move $\Theta(n^{\varepsilon})$ items from one side of the base to the opposite side in $\Theta(n^{\varepsilon/2})$ time by moving them up to a level that is a mesh of size $\Theta(n^{\varepsilon})$, sideways through this mesh, and then down to the base [26].

4.5 Improved pyramids

Many pyramid modifications have been suggested, some of which are surveyed in [27,28]. However, few of the suggestions are directed towards quickly solving perimeter-bound problems, which is the class of problems most amenable to improvement through modest modifications. The goal should be poly-log time since the communication diameter is logarithmic. This implies that individual PEs have only a poly-log number of pixels, so for pyramids the idea of fewer, bigger PEs is of less use. To achieve poly-log time, machines must be able to quickly move $\Theta(n)$ words from one edge of the image to the opposite one. Two suggestions that provide such ability are the orthogonal-trees organization from VLSI [29] and a proposal in [30].

The *orthogonal trees* of size n^2 (also called a mesh of trees) has a base that is a mesh of size n^2. Above each row and each column there is a binary tree, where the trees are disjoint except for the base (see Fig. 5a). Variations include machines where the vertices of the trees are simple message-routing switches, switches with additional logic capabilities, or complete PEs. Such variations do not significantly alter the times in an O-notational analysis.

The proposal in [30], illustrated in Fig. 5(b), is slightly closer to the standard pyramid. To construct it, let n be an integral power of 4, start with the standard pyramid of size n^2, and remove the top $\frac{1}{2}\lg(n)$ levels. The base mesh connections, and all tree connections between levels, are kept. The level above the base is tessellated into squares of 4 PEs, with the standard mesh connections between the squares but "better" connections within each square. The next level is tessellated into squares of 16 PEs, again with the standard mesh connections between the squares and "better" connections within squares. This continues until the top level (the middle of the original pyramid), a single square with "better" connections. Here "better" means any interconnection scheme, such as a crossbar, cube-connected cycles, shuffle connections, etc., such that data in a square of s PEs can be sorted in $O(\log^2(s))$ time. The squares on each level are called "data squares" since they correspond to the "data-square" algorithm approach for standard pyramids [26]. This modified pyramid will be unimaginatively referred to as a *data-square pyramid*.

Algorithms for meshes and pyramids

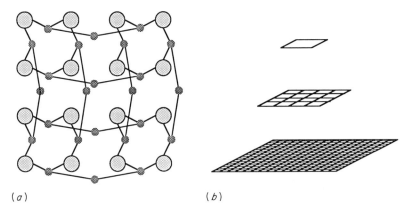

Fig. 5. (a) Orthogonal trees (with mesh connections omitted). (b) Data squares, with 256 base PEs.

The fact that the trees of an orthogonal tree machine are disjoint enables $O(n \star \log(n))$ words to be moved from one base edge to the opposite one in $\Theta(\log(n))$ time. (It is $n \star \log(n)$, and not just n, because the transmission can be pipelined, starting n words at time 1, then n more at time 2, and so on.) On the data-square pyramid, the sorting capability within the data squares allows one to rapidly rearrange data to send it between levels or across the top level, with the result that it too can move $O(n \star \log(n))$ words from one base edge to the opposite one in poly-log time.

Using these rapid data movement capabilities, all of the perimeter-bound problems mentioned can be easily solved in poly-log time, instead of the polynomial time required on the standard pyramid. Further, poly-log time can be attained by straightforward implementations of standard divide-and-conquer algorithms, so the programming effort is far easier than, say, a mesh with broadcasting, while achieving better times.

While the orthogonal trees and data-square pyramid can solve many perimeter-bound problems in poly-log time, they still require $\Omega(n)$ time for all strongly global problems since there are still only $\Theta(n)$ wires crossing a plane through their middles. These machines can quickly move $O(n \star \log(n))$ words, but not $\Omega(n^2)$. They are compromises which solve an important class of problems in poly-log time, without the cost of an interconnection scheme able to quickly solve some strongly global problems.

4.6 Hypercubes

While the emphasis has been on modifications of meshes and pyramids, possibilities more suited to some strongly global problems should also be mentioned. One obvious possibility is to use higher-dimensional meshes, since they have higher bandwidths between halves. Another possibility is the hypercube, an architecture of intense current interest. For a nonnegative integer d, the d-*dimensional* (binary) *hypercube* has 2^d PEs, indexed by the numbers 0 through $2^d - 1$. If the binary expansion of a PE's number is $i_{d-1} \ldots i_1 i_0$ then the PE is connected to the d PEs with binary expansions of the form $i_{d-1} \ldots j_k \ldots i_1 i_0$ for $0 \leq k \leq d-1$, where $j_k = 1 - i_k$. An SIMD hypercube is under construction [31], but most hypercubes and hypercube algorithms are MIMD.

Meshes can be embedded into hypercubes so that neighbours in the mesh are mapped to neighbours in the hypercube. For example, to embed an 8×8 mesh into a 6-dimensional hypercube, the mesh PE at (i,j) would map to the hypercube PE number $(i_2 i_1 i_0 j_2 j_1 j_0)$, where the maps $i \to i_2 i_1 i_0$ and $j \to j_2 j_1 j_0$ are 3-bit Gray codes. Using this embedding, any mesh algorithm can be run at the same speed on a hypercube of the same size. (This assumes that the dimensions of the mesh are powers of 2. A 5×5 mesh does not embed into a 5-dimensional hypercube despite the fact that the hypercube has enough PEs.)

There is no neighbour-preserving mapping of a pyramid of size > 1 into a hypercube since the pyramid has cycles of odd length and hypercubes only have cycles of even length. However, a pyramid of P PEs can be mapped into a hypercube of dimension $\lceil \lg(P) \rceil$ such that neighbouring PEs in the pyramid are mapped to PEs no more than 3 communication links apart in the hypercube. (Any graph can be mapped into a hypercube so that adjacent vertices in the graph are no more than 2 links apart in the hypercube, but the hypercube may need to be much larger than the graph.) Therefore a hypercube can simulate a single communication step of a pyramid in $\Theta(1)$ time, and can run pyramid algorithms almost as fast as the pyramid can. Further, the hypercube can solve several strongly global problems, such as sorting and image rotation, in poly-log time.

Besides quickly simulating meshes and pyramids, and providing fast sorting and general data movement, the hypercube has a recursive structure which naturally implements divide-and-conquer algorithms. However, the wiring is a limiting factor, and for some time it may be impractical to build a hypercube of, say, a million PEs. Further, a large hypercube will cost more than a mesh of the same size.

One way to try to gain the algorithmic advantages of the hypercube while retaining the cost advantages of a mesh is to tie them together. For example, suppose there is a 512×512 mesh of bit-serial PEs (i.e. typical PEs on image-processing meshes), tessellated into 16×16 squares. To each square is attached a more powerful PE, with these powerful PEs interconnected as a

10-dimensional hypercube. Modifying a suggestion in [30], perhaps the memories of the mesh PEs can be arranged so that each bit plane in a square looks like a block of eight 32-bit words to the corresponding hypercube PE, or perhaps as sixteen 16-bit words addressable via rows or columns. The mesh would presumably be SIMD, with either the entire mesh performing a single instruction at a time, or with each square separately controlled by its hypercube PE. The hypercube would be MIMD.

Algorithms for such a machine would use the mesh to perform most mesh-local operations used in initial image processing and the hypercube would perform the more communication-intensive pyramid-local, perimeter-bound and strongly global operations used in later stages of image understanding. The mesh algorithms would draw heavily from algorithms for current image-processing machines, while the hypercube algorithms would draw more from current work on higher-level image understanding and artificial intelligence and would be more experimental.

The hypercube algorithms would almost certainly undergo a long period of experimentation, since so little is known about image understanding. In such a research situation the hypercube has a high bandwidth from any region to any other, so it does not immediately impose communication bottlenecks on problems. It is also sufficiently regular so that algorithm development is fairly straightforward. These features encourage experimentation since algorithms can be developed and run quickly enough to keep up the interest of the investigator.

5 CONCLUSION

This chapter has studied some characteristics of problems and parallel algorithms in order to suggest changes to current image-processing architectures, with the goal of designing architectures with faster execution times on image understanding algorithms to be developed in the future. SIMD/MIMD tradeoffs and communication requirements of various classes of problems were examined, and some suggestions were made to improve deficiencies of current architectures.

The alternatives discussed herein are biased by the author's interest in algorithms, and, while an attempt has been made to consider possibilities that might be cost effective, some may turn out to be unreasonable. On the other hand, unless future machines address some of the points discussed, researchers in image understanding will find that it is too hard to use the machines efficiently, and that the machines are internally communication-bound on many problems. For advanced image-understanding projects, programming costs will far exceed hardware costs, and the machines must be fast enough on a wide range of problems to encourage sustained experimentation. Rapid, repeated experimentation, along with the development of theoretical underpinnings, seems crucial to the development of image understanding.

ACKNOWLEDGMENTS

This research was partially supported by Naval Research Laboratory contract 65-2068-85, and National Science Foundation grant DCR-8507851.

REFERENCES

[1] Winston, P. H. (1984). *Artificial Intelligence*. Addison-Wesley, Reading, Massachusetts.
[2] PPRC (Parallel Processing Research Council) (1985). Responses to first survey of parallel processing projects (compiled by Arvind). *Tech. Rep. PPRC*.
[3] Beyer, W. T. (1969). Recognition of topological invariants by iterative arrays. PhD thesis, MIT.
[4] Dyer, C. R. and Rosenfeld, A. (1977). Cellular pyramid for image analysis. TR-544, *Comp. Sci. Center, Univ. Maryland*.
[5] Levitt, K. N. and Kautz, W. H. (1972). Cellular arrays for the solution of graph problems. *Commun. ACM* **15,** 789–801.
[6] Stout, Q. F. (1982). Using clerks in parallel processing. In *Proc. 23rd IEEE Symp. on Foundations of Computer Science*, pp. 272–279.
[7] von Neumann, J. (1966). *The Theory of Automata: Construction, Reproduction, and Homogeneity* (ed. A. Burks). University of Illinois Press, Urbana.
[8] Dyer, C. R. and Rosenfeld, A. (1981). Parallel image processing by memory augmented cellular automata. *IEEE Trans. PAMI* **3,** 29–41.
[9] Miller, R. and Stout, Q. F. (1985). Geometric algorithms for digitized pictures on a mesh-connected computer. *IEEE Trans. PAMI* **7,** 216–228.
[10] Miller, R. and Stout, Q.F. (1985). Varying diameter and problem size in mesh-connected computers. In *Proc. 1985 Int. Conf. on Parallel Processing*, pp. 697–699.
[11] Nassimi, D. and Sahni, S. (1980). Finding connected components and connected ones on a mesh-connected parallel computer. *SIAM J. Comp.* **9,** 744–757.
[12] Thompson, C. D. and Kung, H. T. (1977). Sorting on a mesh-connected parallel computer. *Commun. ACM* **20,** 263–271.
[13] Uhr, L. (1972) Layered "recognition cone" networks that preprocess, classify, and describe. *IEEE Trans. Comp.* **21,** 758–768.
[14] Uhr, L. (1980). Psychological motivation and underlying concepts. In *Structured Computer Vision: Machine Perception through Hierarchical Computation Structures* (ed. S. Tanimoto and A. Klinger). pp. 1–30. Academic Press, New York.

[15] Frederickson, G. N. (1983). Distributed algorithms for selection in sets. *Tech. Rep., Comp. Sci., Purdue Univ.*
[16] Miller, R. and Stout, Q. F. (1984). Convexity algorithms for pyramid computers. In *Proc. 1984 Int. Conf. on Parallel Processing*, pp. 177–184.
[17] Stout, Q. F. (1985). Pyramid computer solutions of the closest pair problem. *J. Algorithms* **6**, 200–212.
[18] Stout, Q. F. (1985). Tree-based graph algorithms for some parallel computers. In *Proc. 1985 Int. Conf. on Parallel Processing*, pp. 727–730.
[19] Tanimoto, S. L. (1982). Programming techniques for hierarchical parallel image processors. In *Multicomputers and Image Processing Algorithms and Programs* (ed. K. Preston and L. Uhr), pp. 421–429. Academic Press, New York.
[20] Hirschberg, D. S., Chandra, A. K. and Sarwate, D. V. (1979). Computing connected components on parallel computers. *Commun. ACM* **22**, 461–464.
[21] Atallah, M. J. and Kosaraju, S. R. (1984). Graph problems on a mesh-connected parallel processor array. *J. ACM* **31**, 649–667.
[22] Bokhari, S. H. (1984). Finding maximum on an array processor with a global bus. *IEEE Trans. Comp.* **33**, 133–139.
[23] Stout, Q. F. (1982). Broadcasting in mesh-connected computers. In *Proc. 1982 Conf. on Information Science and Systems*, pp. 85–90.
[24] Kumar, V. K. P and Raghavendra, C. S. (1985). Image processing on an enhanced mesh connected computer. *Tech. Rep., Elec. Eng. Systems, Univ. California.*
[25] Stout, Q. F. (1986). Optimal bussing schemes for meshes. In preparation.
[26] Miller, R. and Stout, Q. F. (1986). Data movement techniques for the pyramid computer. *SIAM J. Comp.* In press.
[27] Uhr, L (1983). Pyramid multi-computers, and extensions and augmentations. TR 523, *Comp. Sci., Univ. Wisconsin–Madison.*
[28] Uhr, L. (1984). Augmenting pyramids and arrays by compounding them with networks. In *Proc. Workshop on Algorithm-Guided Parallel Architecture for Automatic Target Recognition*, pp. 317–327.
[29] Ullman, J D. (1984). *Computational Aspects of VLSI.* Computer Science Press, Maryland.
[30] Stout, Q. F. (1984). Mesh and pyramid computers inspired by geometric algorithms. In *Proc. Workshop on Algorithm-Guided Parallel Architectures for Automatic Target Recognition*, pp. 293–315.
[31] TMC (Thinking Machines Corp.) (1985). The Connection Machine[TM] Supercomputer: a natural fit to application needs. *Tech. Rep., TMC.*

Chapter Ten
Pyramid Architectures

D. H. Schaefer

1 INTRODUCTION

Collections of processing elements arranged in a pyramid configuration are gaining increased attention in the computer research community. At least two projects involving hardware versions of pyramids are in progress in the United States, and an ambitious project is underway in Italy.

Pyramid structures differ both in the architecture of their processing elements and in the type and mode of intercommunication between processing elements. When two-dimensional schematics [1,2] are used to represent pyramid designs, differences are highlighted. The PCLA design of Tanimoto and colleagues [3], the PAPIA design of Cantoni et al. [4], and the MPP Pyramid design being pursued by the author and his colleagues [5] will be examined.

2 TWO-DIMENSIONAL NOMENCLATURE

In order to describe the structure of multidimensional computing structures, it is advantageous to utilize components in these representations that are themselves multidimensional. Only single levels of pyramids will be shown; therefore a collection of two-dimensional components, such as presented in Fig. 1, will suffice for these descriptions. Each component accepts a two-dimensional array as input and provides as output an array transformed in some manner. Elements of an array will be either "one" or "zero", as all pyramids to be discussed operate in a bit-serial mode.

The "geometric components" of Fig. 1 are of the greatest interest when describing pyramids, as they provide paths for data transfer from one

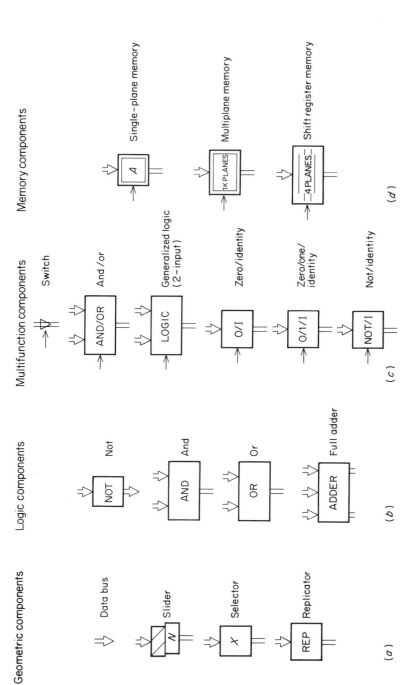

Fig. 1. Glossary of two-dimensional components.

Pyramid architectures 169

element of the pyramid structure to another. The first "component", the data bus, is the equivalent of an identity operator, the output array being identical with the input array. A data bus transferring an n by n array contains n^2 individual data paths. In the case of binary data arrays, these paths are individual wires or perhaps optical fibres.

The slider component transforms the input array by translating the input one element in the direction indicated by the letter in the lower box of the symbol. For instance, the slider of Fig. 1 translates the input array one element north as shown in Fig. 2. Zeros fill in elements on an edge that are undefined during the operation. An alternative use of a slider is for data input when edge elements acquire values fed from an external source.

Selector and replicator components are distinguished from data busses and sliders in that the dimension of the output array is different from that of the input array. An X-selector, for instance, outputs only the X-elements of its input, as shown in Fig. 3. V-, Y- and Z-selectors output only the V-, Y- and Z-elements in a similar fashion. Therefore the output of a selector contains only one quarter as many elements as the input array. In contrast, a replicator increases the dimension of the input array by providing four identical outputs for each input element as shown in Fig. 4.

Selectors and replicators provide the means to transfer information from one level of a pyramid to the next, while sliders provide intra-level communication. It should be noted that geometric components contain only wires. Their existence is often not recognized as they are simply part of the wiring.

V_1	X_1	V_2	X_2	V_3	X_3	V_4	X_4
Y_1	Z_1	Y_2	Z_2	Y_3	Z_3	Y_4	Z_4
V_5	X_5	V_6	X_6	V_7	X_7	V_8	X_8
Y_5	Z_5	Y_6	Z_6	Y_7	Z_7	Y_8	Z_8
V_9	X_9	V_{10}	X_{10}	V_{11}	X_{11}	V_{12}	X_{12}
Y_9	Z_9	Y_{10}	Z_{10}	Y_{11}	Z_{11}	Y_{12}	Z_{12}
V_{13}	X_{13}	V_{14}	X_{14}	V_{15}	X_{15}	V_{16}	X_{16}
Y_{13}	Z_{13}	Y_{14}	Z_{14}	Y_{15}	Z_{15}	Y_{16}	Z_{16}

(a)

Y_1	Z_1	Y_2	Z_2	Y_3	Z_3	Y_4	Z_4
V_5	X_5	V_6	X_6	V_7	X_7	V_8	X_8
Y_5	Z_5	Y_6	Z_6	Y_7	Z_7	Y_8	Z_8
V_9	X_9	V_{10}	X_{10}	V_{11}	X_{11}	V_{12}	X_{12}
Y_9	Z_9	Y_{10}	Z_{10}	Y_{11}	Z_{11}	Y_{12}	Z_{12}
V_{13}	X_{13}	V_{14}	X_{14}	V_{15}	X_{15}	V_{16}	X_{16}
Y_{13}	Z_{13}	Y_{14}	Z_{14}	Y_{15}	Z_{15}	Y_{16}	Z_{16}
0	0	0	0	0	0	0	0

(b)

Fig. 2. North slider: (a) input; (b) output.

V_1	X_1	V_2	X_2	V_3	X_3	V_4	X_4
Y_1	Z_1	Y_2	Z_2	Y_3	Z_3	Y_4	Z_4
V_5	X_5	V_6	X_6	V_7	X_7	V_8	X_8
Y_5	Z_5	Y_6	Z_6	Y_7	Z_7	Y_8	Z_8
V_9	X_9	V_{10}	X_{10}	V_{11}	X_{11}	V_{12}	X_{12}
Y_9	Z_9	Y_{10}	Z_{10}	Y_{11}	Z_{11}	Y_{12}	Z_{12}
V_{13}	X_{13}	V_{14}	X_{14}	V_{15}	X_{15}	V_{16}	X_{16}
Y_{13}	Z_{13}	Y_{14}	Z_{14}	Y_{15}	Z_{15}	Y_{16}	Z_{16}

(a)

X_1	X_2	X_3	X_4
X_5	X_6	X_7	X_8
X_9	X_{10}	X_{11}	X_{12}
X_{13}	X_{14}	X_{15}	X_{16}

(b)

Fig. 3. *X-Selector: (a) input; (b) output.*

X_1	X_2	X_3	X_4
X_5	X_6	X_7	X_8
X_9	X_{10}	X_{11}	X_{12}
X_{13}	X_{14}	X_{15}	X_{16}

(a)

X_1	X_1	X_2	X_2	X_3	X_3	X_4	X_4
X_1	X_1	X_2	X_2	X_3	X_3	X_4	X_4
X_5	X_5	X_6	X_6	X_7	X_7	X_8	X_8
X_5	X_5	X_6	X_6	X_7	X_7	X_8	X_8
X_9	X_9	X_{10}	X_{10}	X_{11}	X_{11}	X_{12}	X_{12}
X_9	X_9	X_{10}	X_{10}	X_{11}	X_{11}	X_{12}	X_{12}
X_{13}	X_{13}	X_{14}	X_{14}	X_{15}	X_{15}	X_{16}	X_{16}
X_{13}	X_{13}	X_{14}	X_{14}	X_{15}	X_{15}	X_{16}	X_{16}

(b)

Fig. 4. *Replicator: (a) input; (b) output.*

Two-dimensional logic components are arrays of active devices that perform logical functions over an array. For instance, the NOT component (Fig. 1) outputs an array where every element is the complement of its input. The AND and OR components of Fig. 1 each receive two arrays as input. However, these devices can be generalized to receive any number of array inputs.

Pyramid architectures

Multifunction component operation can be altered by a control signal. For instance, the AND/OR component (Fig. 1) can be programmed to perform either the AND or the OR function, depending on the control-signal input. The arrow on the left is the control input. This control is replicated and these signals in turn control all the individual circuits in the two-dimensional component. An important multifunction operator is the LOGIC component that can produce any of the Boolean functions of its inputs. The ZERO/IDENTITY component either provides a field of zeros or transmits the input unaltered, while the NOT/IDENTITY component, in a similar manner, either provides the complement of the input, or transmits this input without change. The NOT/IDENTITY can alternately be thought of as a device that matches the input-element values to that of the control-signal value, outputing ones at those locations where the value of the input element and the value of the control agree, and outputing zeros where they disagree.

Memory of a single binary array is provided by the single-plane memory (or "register array"), which is an array of flip-flops. In a similar manner, a multiplane memory ("memory array") is an array of random-access memories (each RAM storing 1024 bits in Fig. 1), while the shift register memory is an array of shift registers (four-stage shift registers in the Fig. 1 example).

3 REPRESENTATION OF THE SUM-OR

An example of the use of two-dimensional symbolism is the representation of a circuit whose output is the OR of all its input elements. Fig. 5 represents

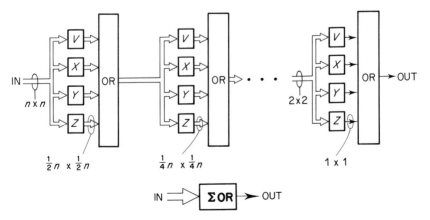

Fig. 5. *SUM-OR circuit represented as a pyramid.*

such a "SUM-OR" where the previously defined geometric and logical components are utilized. This diagram exhibits the SUM-OR circuit as a very primitive pyramid.

4 TWO THEORETICAL PYRAMIDS

The SUM-OR is a fixed piece of hardware that performs only one function. At the opposite end of the flexibility curve is a generalized representation of a two-dimensional non-overlapped quaternary tree pyramid containing one register array at each level (Fig. 6). The value of any element of a register array can be replaced (in one cycle) by a logical function of itself, its eight neighbours on the same level as itself (its siblings), its parent on the level above, and its four children on the level below. Using Tanimoto's nomenclature [6], the value of each element can be replaced by any Boolean function of its fourteen-element "pyramidal neighbourhood". The particular function will be the same for all elements on a given level.

Control signals into the fourteen input LOGIC components determine which of the $2^{2^{14}}$ functions is generated. The number of *control* lines needed to produce $2^{2^{14}}$ different Boolean functions is 2^{14}, or 16 384 control lines! One cannot design pyramids with logic providing all the millions of millions of functions of the 14 inputs (or even more for overlapped pyramids), but instead must compromise on the number of available functions that can be executed in one cycle.

In a non-overlapped pyramid each element of an array has a direct connection to only one parent. If the connections to children are as shown in Fig. 7, each element has a direct connection to two parents. This particular overlapped configuration can sense (in one cycle) the presence of a given pattern in *any* 2×2 block on the level below. In contrast, the non-overlapped configuration can only sense (in one cycle) the presence of given patterns in the 2×2 blocks associated with each parent (i.e. a V, X, Y, Z block of Fig. 2*a*.

5 THREE PYRAMID DESIGNS

Insight into the compromises that must be made when designing pyramid structures is provided by examination of three designs that are in various stages of hardware implementation.

The Pyramid Cellular Logic Array (PCLA) (Fig. 8) has been designed at the University of Washington, and prototype integrated circuits have been fabricated [6]. The PCLA design consists of a "matching circuit" (top

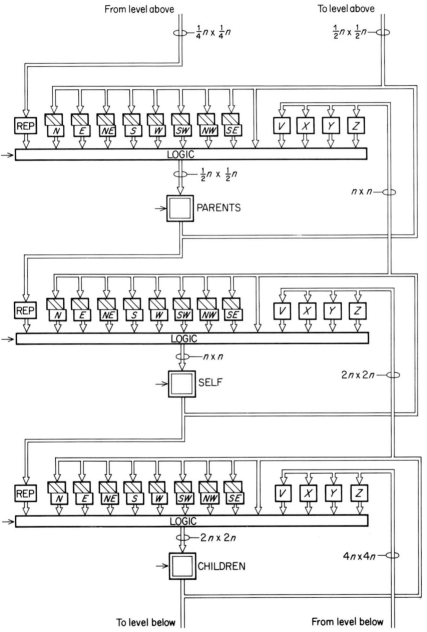

Fig. 6. Generalized pyramid representation.

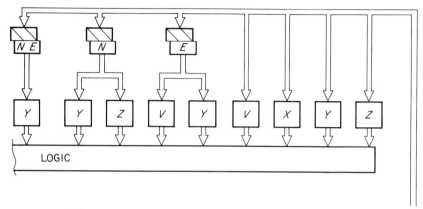

Fig. 7. Connections to children for an overlapped pyramid.

portion of the figure), three register arrays and a memory array at each level. The matching circuit produces either an "AND MATCH" or an "OR MATCH". The matching is with control signals to the NOT/I and O/1/I components. Control to the O/1/I components determines whether the given neighbour should enter into the match or become a "don't care". Control to the NOT/I components determines if the match desired for the given neighbour is a "one" or a "zero". The results of the fifteen match operations can either be ANDed or ORed, depending on the control to the AND/OR component, and then loaded into either the "propagation" array P, the "local" array L, or the "gating" array G. (G is used rather than University of Washington's C for uniformity in this article.) For example, to load the P-registers with the AND of their children (i.e. to calculate a match where all children are "ones" and the value of parent, siblings and the L-register is of no interest), the AND/OR is set to "AND", the NOT/I components attached to the V-, X-, Y- and Z-components are set to "I", while the O/1/I components beneath them are also set to "I". All other O/1/I components are set to "1" to place them in a "don't care" state.

The G-register is a mask inhibiting clock signals to any element whose value in the G-array is zero. Therefore, at these locations, no change in the value of elements occur. The matching circuit can produce many of the most used functions of its fifteen inputs. To perform the often used exclusive OR operation, however, the mask register must be used in conjunction with the matching circuit, requiring two cycles to perform this function.

Another pyramid, Italy's Pyramid Architecture for Parallel Image Analysis (PAPIA) is a joint project of a large number of Italian organizations. Its design (Fig. 9) embodies a simpler interlevel connection scheme than the PCLA.

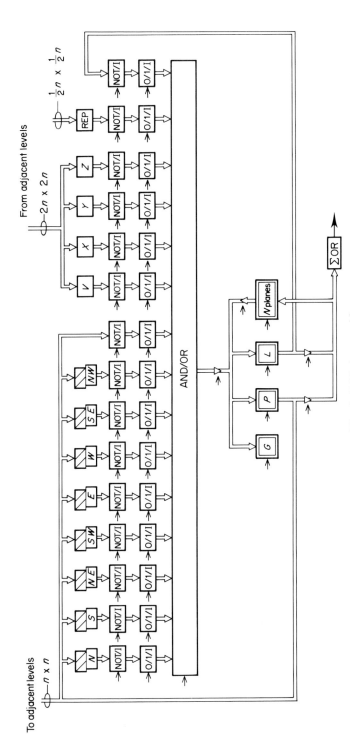

Fig. 8. PCLA pyramid.

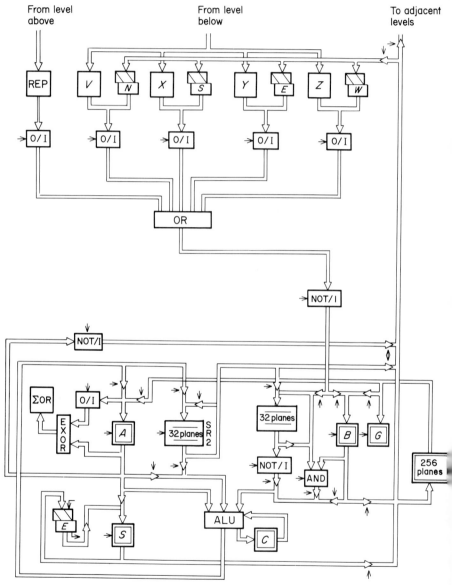

Fig. 9. PAPIA pyramid.

Pyramid architectures

The designers of the PAPIA made the decision to limit the neighbourhood that is accessible in one cycle to either (i) children and parent or (ii) nearest-neighbour siblings (i.e. excluding those on the diagonal). The OR and NOR of these inputs are available for input by memory planes A and B and the shift registers. The input to the ALU can be either the OR, AND, NOR or NAND of the pyramid neighbours of A; programming of the two NOT/I components provides the four functions.

The G-plane is the mask array, while the S-plane is used for input and output of data. The length of the shift register memory planes can be varied.

The third design, the Massively Parallel Processor Pyramid (MPP Pyramid) utilizes custom integrated circuits designed for the Massively Parallel Processor [7]. Other circuitry has been combined with these custom chips to form the pyramid structure shown in Figure 10. A three-level version is in operation and a five-level unit is expected to be functional early in 1986.

Communication with siblings is accomplished by loading the P-register array through a slider with neighbouring elements. This slid data is then combined with the data stored in memory by use of the LOGIC component. During transfer of data between levels, children and parents move from the data bus on their own level through a tristate switch array to the data bus of the receiving level. Usually data then flows through the LOGIC component into the P-register array. To load register array P with any function of all of its fourteen pyramidal neighbours requires at least 28 cycles, 23 of these cycles being used to form the function of the element and its eight siblings.

The LOGIC component is relieved of many tasks by the full adder circuitry shown in the lower portion of Fig. 10. Addition can take place simultaneously with sliding operations. The mask array G can mask either arithmetic operations or logic operations, or both. The shift register array length can be varied.

The MPP Pyramid has been programmed for object recognition, object counting, and segmentation algorithms. In these programs the most used function involving neighbours is the AND of children, most generally the AND of three out of the four children. Next is the AND and OR of an element and its eight siblings, followed by arithmetic summing of the values of children. At no time has there been a need to produce any functions combining children and siblings. The only instance of a function of a parent and another element is masked loading of children from parents. The use of sliding operators to compensate for lack of overlapping is common.

6 CONCLUDING REMARKS

Pyramid architectures must provide a wide variety of different operations.

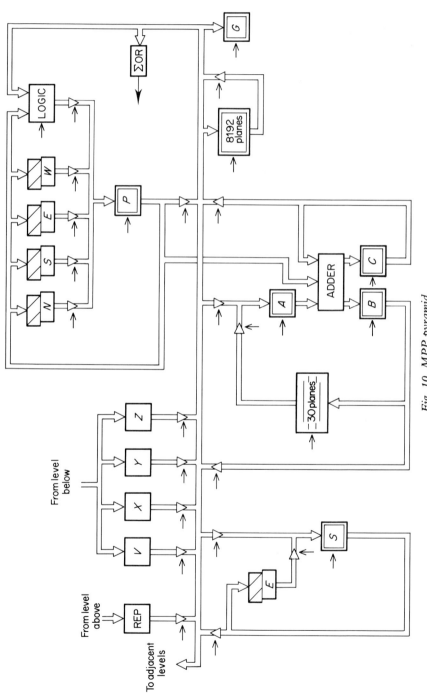

Fig. 10. MPP pyramid.

Pyramid architectures 179

Two important tasks are to be able to quickly (i) compute a large range of functions of either siblings or children in one cycle, and (ii) produce and efficiently store the sum of children. The PCLA performs excellently for task (i), and very poorly for task (ii). The MPP Pyramid, on the other hand, does well for task (ii) and very poorly for task (i). The PAPIA steers a middle ground with a communication network that quickly provides a limited number of either sibling or children functions, and also provides logic for addition. None of the designs incorporates any overlapping, a definite deficiency.

This chapter has contrasted pyramid designs by use of two-dimensional representations. Although in the pyramids discussed every level was identical, the two-dimensional representation can be used to represent future pyramids that may have different architectures at each level.

ACKNOWLEDGMENTS

I wish to acknowledge the contributions of my students, Mrs Ping Ho, Mr Cesar Vallejos and Mr Gregory Wilcox, and the editorial assistance of my wife, Maxine.

REFERENCES

[1] Schaefer, D. H. (1982). Spatially parallel architectures. An overview. *Computer Design* (August), pp. 117–124.
[2] Schaefer, D. H. (1985). Representations of spatially parallel architectures. In *Computer Architectures for Spatially Distributed Data* (ed. H. Freeman & G. Pieroni). pp. 57–74. Springer–Verlag, Berlin.
[3] Tanimoto, S. L. (1983). A pyramidal approach to parallel processing. In *Proc. 10th Ann. Int. Symp. on Computer Architecture, Stockholm, June*, pp. 372–378.
[4] Cantoni, V., Ferretti, M., Levialdi, S. and Maloberti, F. (1985). A pyramid project using integrated technology. In *Integrated Technology for Parallel Image Processing* (ed. S. Levialdi), pp. 121–132. Academic Press, London.
[5] Schaefer, D. H., Wilcox, G. C. and Harris, W. J. (1985). A pyramid of MPP processing elements—experience and plans. In *Proc. 18th Ann. Hawaii Int. Conf. on System Sciences*, Vol. 1, pp. 178–184.
[6] Ligocki, T. J. (1984). The hierarchical cellular logic (HCL) chip. Masters Project Report, Dept Comp. Sci. Univ. Washington.
[7] Burkley, J. T. (1985). MPP VLSI multiprocessor integrated circuit design. In *The Massively Parallel Processor* (ed. J. Potter) pp. 205–215. MIT Press, Cambridge, Massachusetts.

Chapter Eleven
Contour Labelling by Pyramidal Processing

V. Cantoni and S. Levialdi

1 INTRODUCTION

Many different strategies have been applied to the analysis and description of contours of digital images in the last twenty years (see e.g. [1–3]) since such contours probably contain the most significant information [4] for compact storage, efficient recognition and fast display of the shapes contained in the considered images.

Most algorithms for shape description and measurement make use of sequential procedures (since basically all image-processing centres use sequential computers). These, after suitable preprocessing (i.e. cleaning up the image and removing noise as much as possible), detect a number of critical points and salient features which reveal either points of inflection or sharp concavities (convexities), so pointing at contour elements of particular relevance. Since concavities (convexities) are not diameter-limited properties [5], it is in general difficult to establish an *a priori* neighbourhood size in order to ensure that all such concavities (convexities) will be detected. A number of different approaches have been attempted, including the minimum-perimeter polygon (MPP) [6], the convex hull [7], the concavity tree [8], in order to tackle and solve such problems by means of an analogue model such as an elastic rubber band around the digital object, the minimum box containing the object or a graph-symbolic structure that associated each node with a concavity region.

With the advent of parallel machines (also known as multicomputers) [9], new algorithms have been proposed in order to exploit the concurrent processing due to the simultaneous execution of an instruction (or of different instructions, according to the particular operating mode of the system) in a collection of processors, each one of which can have access to all the information regarding the neighbourhood processors (pixel elements) belonging to the image. Most parallel machines designed and built for image-processing applications operate in SIMD mode and are therefore able to efficiently compute local operations on the whole image in a few clock cycles.

Typically three by three neighbourhoods are considered (for reasons of hardware simplicity and large-scale integration constraints) and, if larger neighbourhoods are required, propagation of the information is performed, resulting in a slight time overhead.

The available methods usually require special operations in order to take into account the contextual information for contour evaluation. On the other hand SIMD machines can simultaneously broadcast information to neighbours (which can both send and receive data) in a natural way. Owing to this particular feature, it is relatively easy to model growth processes [10], genetic laws (like Conway's Life Game for instance), region-growing operators, etc.

2 METHOD

In a preliminary report, Skliar and Loew [11], have suggested a new method for the characterization of shape based on the analogy with a diffusion process acting on a physical object having the same shape as the digital object on which the contour will be labelled.

Their method starts by assigning to all the contour elements of the digital object to be labelled, an arbitrary integer value which initializes the diffusion process. They then assume that a standard heat-diffusion process is originated on the contour, described by a finite-difference equation (an approximation to the equation using partial derivatives), and that this process propagates towards the interior of the object from each element to its neighbours (and vice versa), as formally described by the following expression:

$$f_{t+1}(i,j) = f_t(i,j) + D[(f_t(i-1,j) + f_t(i+1,j) + f_t(i,j-1) + f_t(i,j+1)) - N(i,j) f_t(i,j)].$$

At each iteration of the diffusion step, new pixels will be contaminated provided they lie in the interior of the object. After a number of steps the contour elements will preserve high values corresponding to local convexities and will produce a sharp decrement in value corresponding to local concavities.

Contour labelling by pyramidal processing

$f_{t+1}(i,j)$ represent the value of the pixel at position (i,j) at time $t+1$. D is the diffusion coefficient, having a value between 0 and the D_{max} due to the diffusion condition which splits the local value of each pixel content among all the defined neighbours; since the number of neighbours in 4-connectivity is four, D_{max} is 0.25. If $D > D_{max}$ then negative values (without physical meaning) may appear for some pixels, and instability of the process may occur. $N(i,j)$ is the number of neighbouring elements belonging to the object (using 4-connectivity, this number may range between 0 and 4).

In the initial state only contour elements will differ from 0; after the first iteration both contour elements and their neighbours (interior to the object) will be contaminated ("warmed up") and so on, reproducing the diffusion process from the contour towards the inside of the object. During this process the values are distributed along the contour depending on the local shape configuration; these values may be used for contour labelling.

Although the diffusion process is strictly isotropic, only four directions

Fig. 1. Original object: a frame.

are considered, namely the north, south, west and east, corresponding to the 4-connectivity [12] mode of digital geometry.

As mentioned in Section 1, analogies of physical processes have been used for developing algorithms in image processing on many occasions. For instance, the skeleton transformation [13] originated from a burning wavefront propagating in a windless atmosphere, generating, at the fire extinction points, skeleton elements which are normally used for a perceptual characterization of the original object; this is particularly effective whenever elongated shapes are taken into account.

3 SIMULATED DIFFUSION PROCESS ON AN ARRAY

In order to visualize the diffusion method applied to a digital binary image, let us consider a specific object, a "frame", which is substantially made of a digital square having a hollow square in its centre (see Fig. 1). This square

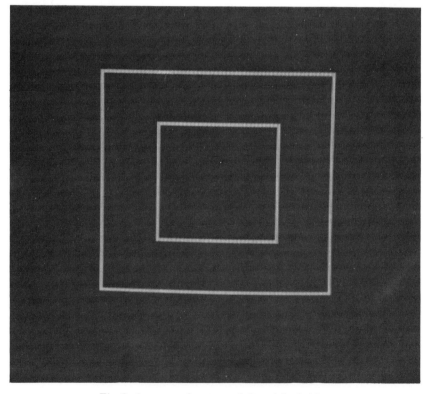

Fig. 2. 4-connected contour of the original object.

Contour labelling by pyramidal processing

contains four external corners (convexity points) and four internal corners (concavity points). Figure 2 shows the 4-connected contour of the "frame". We shall now assume that the diffusion coefficient D is 0.25 (for maximum speed), and for 10 iterations we obtain the results shown in Fig. 3 (for 20 iterations see Fig. 4 and for 30 iterations see Fig. 5). A uniform grey scale was chosen to show the different values obtained on the border: white represents a high value and black a low one.

The equation given in the previous section is ideally suited for implementation on the family of cellular machines. In fact, the diffusion process is strictly local and may be programmed on SIMD machines so that at each iteration step the diffusion process (within a three by three neighbourhood) will be executed. The diffusion process starting from an initial condition that does not discriminate concavities (convexities) will proceed until minimum (maximum) values are obtained after a number of steps

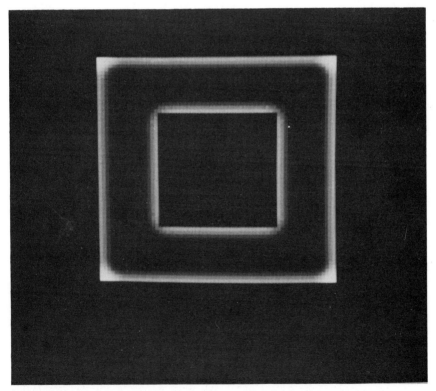

Fig. 3. Result of applying 10 iterations of the diffusion process with $D = 0.25$; a linear grey scale was used to represent the different pixel values.

Fig. 4. Result of applying 20 iterations on the same original object.

(depending on the smoothness of the curvature); a steady state of the system will then be reached in which, again, a uniform value for all pixels is achieved.

The values obtained on the contour of the shape after a given number of iterations may be plotted against the curvilinear length along the contour. In this way the peaks (valleys) of the graph will show the position of convexities (concavities). For a given number of iterations a set of thresholds may be chosen so as to associate values exceeding such thresholds with contour labels.

Figure 6 shows four shapes of varying corner angles of plane silhouettes; the result of 8 iterations of the diffusion process with diffusion coefficient D of 0.25 is shown in Figure 7. We may note that the sharp convex angle at the front of the airplane is the brightest point, the right convex angles are still bright, while the right concave angles are dark.

Contour labelling by pyramidal processing 187

Fig. 5. *Result of applying 30 iterations on the same original object.*

4 SIMULATED DIFFUSION PROCESS ON A PYRAMID

Since concavities and convexities may have a wide range of average curvature for a given diffusion coefficient, and the number of iteration steps producing a meaningful value indicating a concavity (or a convexity) will vary within a wide range (a small number of iterations for a local concavity (convexity) and a large number for a smooth concavity (convexity)). For this reason the different concavities (convexities) will be detected after a different number of iteration steps. In general this number is difficult to determine *a priori*, so that the best architecture for extracting properties of unknown size is a multiresolution one where the processors are organized along a set of tapered parallel planes (each one being a cellular array) and where the overall structure may be considered to be a pyramid—hence the name of this family of multicomputers: pyramid machines. In fact this

Fig. 6. Contours of different plane silhouettes.

architecture may simultaneously analyse and detect concavities (convexities) at different resolution planes in a given number of iterations. High values obtained in each plane indicate the presence of such convexities (low values for concavities). For the above reasons we have chosen the diffusion approach as an interesting scheme for contour analysis and labelling using a pyramid machine such as PAPIA [14].

A given curvature is defined, on the analogue plane, by $1/R^2$, where R is the radius of the osculating circle; since each plane of the pyramid has a scale factor of 2 the same curve will be seen as having curvature radius scaled by 4.

For a given number of steps the diffusion process will cover an equivalent distance between a pixel and its neighbours, then for a pyramid plane revealing a radius of curvature R the approximate number of steps that will maximize the curvature detection will be R, provided the border effects and digitization noise are neglected.

In practice, a digital figure will be present on all planes, and for a given diffusion step number k each plane i will detect, in an optimal way, a curvature equal to $(1/k)^{2+i}$.

Contour labelling by pyramidal processing

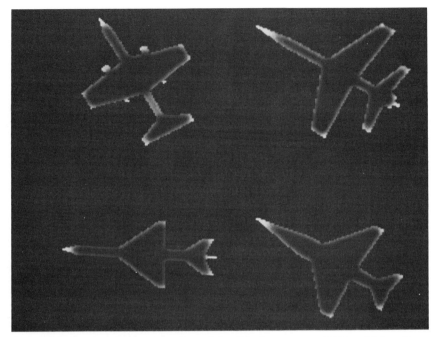

Fig. 7. Results of applying 8 iterations of the diffusion process with $D = 0.25$; a linear grey scale was used to represent the different pixel values.

In conclusion, we may say that the diffusion approach for curvature contour labelling seems well matched to the pyramid family of multiprocessor machines where the SIMD mode is chosen for the whole architecture and the number of iteration steps will equal the minimum curvature radius that is to be detected: this holds for the base plane, and the other planes will be matched to curvatures of ratio 4 according to a geometric series.

REFERENCES

[1] Rosenfeld, A. (1969). Picture processing by computer. *Comp. Surv. ACM* **1**, 147–176.
[2] Davis, L. S. (1976). A survey of edge detection techniques. *Comp. Graph. Image Process.* **4**, 248–270.
[3] Levialdi, S. (1981). Finding the edge. *Proc. NATO ASI on Digital Image Processing* (ed. J. C. Simon and R. M. Haralick), pp. 105–148. Reidel, Dordrecht.
[4] Zusne, L. (1970). *Visual Perception of Form*. Academic Press, New York.

[5] Minsky, M. and Papert, S. (1969). *Perceptrons: An Introduction to Computational Geometry.* MIT Press, Cambridge, Massachussetts.
[6] Sklansky, J., Chazin, R. L. and Hansen, B. J. (1972). Minimum-perimeter polygons of digitized silhouettes. *IEEE Trans. Comp.* **21**, 260–268.
[7] Bass, L. J. and Schubert, S. R. (1967). On finding the disc of minimum radius containing a given set of points. *Math. Comp.* **21**, 712–714.
[8] Sklansky, J. (1972). Measuring concavity on a rectangular mosaic. *IEEE Trans. Comp.* **21**, 1355–1364.
[9] Preston, K., Duff, M. J. B., Levialdi, S., Norgren, P. E. and Toriwaki, J. I. (1979). Basics of cellular logic with some applications in medical image processing. *Proc. IEEE* **67**, 826–856.
[10] Zamperoni, P. (1985). A model for the description of region-growing image operators. *Braunschweig Tech. Univ. Internal Rep.*
[11] Skliar, O. and Loew, M. H. (1985). A new method for characterization of shape. *Patt. Recog. Lett.* **3**, 335–341.
[12] Rosenfeld, A. and Pfaltz, J.L. (1966). Sequential operations in digital picture processing. *J. Ass. Comp. Mach.* **13**, 471–494.
[13] Blum, H. (1967). A transformation for extracting new descriptors of shape. In *Models for the Perception of Speech and Visual Form* (ed. W. Wathen-Dunn), pp. 362–380. MIT Press, Cambridge, Massachussetts.
[14] Cantoni, V., Ferretti, M., Levialdi, S. and Maloberti, F. (1985) A pyramid project using integrated technology. In *Integrated Technology for Parallel Image Processing* (ed. S. Levialdi), pp. 121–132. Academic Press, London.

SYSTEM STUDIES

After a multiprocessor system has been proposed or even constructed, there is still much thought required before the system can be put to good use as an image computer. In this section the first four chapters look at specific systems and suggest how they might be put to work. Tony Reeves and Chang Ho Jeon analyse "Multicluster", an MIMD system proposal for which a small testbed has been assembled, and address in particular the problem of task distribution amongst the available processors. H. J. Siegel and James Kuehn describe the PASM (partitionable SIMD/MIMD) parallel-processing system now being developed at Purdue University, and show how its multiprocessor architecture can be used both to emulate other more specialized architectures and, because of its inherent flexibility, as an effective structure for implementing algorithms that involve widely differing data types and processes. Jean-Luc Basille and his colleagues make the case for a line processor (such as is included in their system SYMPATI) as being a suitable architecture to deal with both iconic and symbolic processes, that is, as an intermediate-level processor. The use of co-occurrence matrices for texture discrimination is well known, but Alain Favre discusses how the VAP processor with its cascade of processors based on look-up tables can be used to generate the matrices and, using a Bayesian approach based on comparison between a manual image segmentation and the measured pixel statistics, VAP makes good use of all its processors in automatically segmenting a liver-cell electron micrograph.

Processing discussed in this section is not exclusively at the intermediate level, but most of the ideas expressed are very largely level-independent. The underlying theme has been to look closely at the processor structure and to look equally closely at the task to be performed. Only by finding how to match one to the other can there be any hope of efficient processing. Barry Gilbert introduces a new dimension for examination: the choice of integrated-circuit technology. He points out that merely employing faster semiconductors does not give a proportionately faster processing speed; it is not always easy or even possible to take full advantage of high gate speeds unless a completely new approach to circuit design is undertaken. This might be expected but it is less obvious that the new technologies (especially gallium arsenide) will only be worthwhile if algorithms are changed as well as the circuit layout.

Chapter Twelve
Computer-Vision Task Distribution on a Multicluster MIMD System

A. P. Reeves and C. H. Jeon

1 INTRODUCTION

A major problem in using parallel systems for a single task is to decompose the task into subtasks that effectively utilize the parallel resources. There are two main types of parallel systems: SIMD, in which a number of processing elements (ALUs with local memory) are controlled by a single programmed control unit; and MIMD, in which a number of independent processors are connected to form a single system. Several highly parallel SIMD systems have been motivated by low-level image-processing applications since there is a good match between image-filtering operations and mesh-connected arrays of processing elements. Highly parallel MIMD systems are now being considered for computer-vision applications. They are much more flexible than SIMD systems and are suitable for both high- and low-level computer vision tasks. The high complexity and flexibility of MIMD systems makes the decomposition and scheduling of a problem much more complex than for the SIMD case. Some tasking strategies for a general MIMD organization called Multicluster are considered in this chapter.

Multicluster [1] is a general MIMD framework for high-speed parallel computation. A system within this framework consists of a set of computational modules called clusters; each cluster contains one or more 32-bit microprocessors, high-speed arithmetic support, local memory and an intercluster interface. The processor interconnection scheme will involve a hierarchy of at least two levels: the intracluster network, which will be determined by technology considerations; and the intercluster network, which will be based more on algorithm and general architecture considerations. The system should be fault-tolerant to failure of one or more clusters (or intercluster connections) and should suffer a graceful degradation of performance as clusters or parts of clusters fail. Furthermore, it should be possible to add new processors to a system as the demands made upon the system are increased.

A Multicluster system is suitable for the parallel implementation of all computer-vision algorithms. Special-purpose SIMD processors may be attached to clusters for more cost-effective low-level image processing and the implementation of other important well-structured algorithms. In a general sense, Multicluster embodies a hierarchical organization of three fundamental parallel-processing architecture types: loosely coupled MIMD, tightly coupled MIMD and SIMD.

1.1 Task decomposition

One of the most important and most difficult aspects of programming a MIMD system for scientific applications is task decomposition. That is, how to divide the given task into a number of subtasks so that the system resources are efficiently utilized in parallel. On Multicluster this may be considered as a two-tier problem. The first part is to determine how to partition the task among the different clusters, and the second part is further decomposing the task to take advantage of multiple processors within a cluster.

Task decomposition on an SIMD system is, in general, not a difficult problem. For example, for low-level image processing there are usually more data elements (pixels) to be processed than there are processors available. Therefore a strategy to allocate a number of pixels to each processor must be devised. Typically there are only a small number of useful strategies that are possible, and each of these may be statically evaluated for optimal performance.

For SIMD-like tasks, task decomposition for the MIMD clusters may be considered in a similar manner. For image-processing applications, for example, it is usually effective to allocate submatrices of consecutive elements to the different clusters for processing owing to the locality property of image-processing algorithms. A lumped model may be used to represent a cluster as a single processing resource with a given performance. This approach is taken in [2] and [3], in which fault tolerance to failing clusters is considered for near-neighbour tasks. For non-deterministic tasks, the situation is much more complex, and task-decomposition strategies need to be developed.

Multicluster MIMD system

Within a cluster there is a true shared memory environment. Effective task decomposition depends upon a number of task and system parameters; this will be the main topic in this chapter. The optimal task-decomposition strategy depends upon a large number of system parameters. Some of the main hardware parameters for a shared memory MIMD system are as follows:

(i) the number of processors;
(ii) memory access time;
(iii) time for an arithmetic operation;
(iv) time for a procedure call;
(v) time for a task rendezvous;
(vi) time for task creation;
(vii) cache memory speed and organization.

In addition, a number of system parameters are also relevant:

(i) memory management policy;
(ii) virtual memory scheme;
(iii) task scheduling scheme.

The memory management policy may be static or dynamic. A static scheme involves no overhead but limits the flexibility of the user programs. For a dynamic memory management scheme the following functions must be implemented: memory allocation, explicit memory deallocation, garbage collection and memory compaction. When there is insufficient primary memory for a task, a virtual memory scheme for swapping data to an auxiliary memory system such as a disk is used. Finally, when there are multiple processors or when time slicing is used, a task scheduling scheme for assigning tasks to processors is necessary.

1.2 MIMD and SIMD systems

SIMD systems have frequently been used for low-level image-processing applications. A very high utilization of processing hardware is possible for these tasks. MIMD systems have been less popular for image-processing tasks; the main reasons in the past have been that the higher functional complexity is too costly and that an SIMD organization is efficient for most applications. The MIMD organization is becoming much more cost-effective with advances in VLSI technology and new architecture designs which include interprocess communication features.

An important advantage of MIMD systems is in the area of fault tolerance. For SIMD systems large amounts of additional hardware are required to ensure correct operation when components fail [4,5]. For MIMD systems there is much more flexibility to deal with faults. Techniques have been developed for distributed systems in which a loss of performance proportional to the size of the contribution of the failed component is experienced and no special hardware to deal with faulty systems is necessary.

A major difference between SIMD and MIMD systems is the programming of applications. Programming SIMD systems is simple and well understood. SIMD programming languages have primitives that are similar to matrix-algebra operators, and the programmer can efficiently design a program based on these primitives. Currently there are no simple primitives for the MIMD case.

While the programming of MIMD systems is very complex, important developments in the specification of computation and the organization of scientific programs may be anticipated in this area. SIMD programming, while it is simple, places a straitjacket on the nature of algorithms that we consider. In the MIMD environment operations do not have to be performed in lock-step to be efficient. More complex operations may be applied to parts of an image that require them.

Much research is needed to develop techniques for programming both tightly and loosely coupled systems. There are similarities between loosely coupled networks and microprocessor local networks. However, there is a major difference in the design goals of the two types of system. In a computer network the usual goal is to share a number of resources among a large number of users. The key here is that the work load is typically a very large number of relatively independent tasks. Frequently, non-numeric applications are important, such as word processing and database management. On the other hand, the objective of the systems considered here is to improve the throughput performance for a single scientific task. Therefore, while similarities occur in the appearance of the hardware organization, there are likely to be significant differences in the details of the implementation of the two system types.

2 MULTICLUSTER

The general design goal of Multicluster is to achieve a cost-effective high-speed MIMD system for scientific applications. There are many formidable problems in designing an MIMD system, including subtask allocation strategy and interconnection network design. However, a set of "ideal" characteristics for a general-purpose MIMD system may be defined which include the following.

Multicluster MIMD system 197

(i) *Extensibility*. It should be possible to add new processors to a system as the demands made upon the system are increased. This is similar in concept to the current facility of most processors that additional memory can be easily added as required.

(ii) *Fault tolerance*. It should be possible to reconfigure the system if any component becomes faulty, so that computation may continue with a minimum of interruption. Furthermore, the reduction in performance should be proportional to the contribution made by the failed processor, and no recompilation of currently executing tasks should be necessary.

(iii) *Programmability*. The user of the system should be able to write programs in a high-level language without having to have a detailed knowledge of the architecture of the system or even knowing how many processors the system has. Programs developed in this language should be both dynamically extensive and fault-tolerant to take advantage of the features mentioned above.

2.1 Multicluster architecture

The general framework of the Multicluster system is outlined in this section. The system may be considered as a set of computing modules or processor clusters (PCs) interconnected by a global interconnection network (GIN) as illustrated in Fig. 1. A cluster consists of four classes of devices connected to a local interconnection network (LIN). A serial general data processor (GDP) and memory module (M) together with the LIN form the heart of the PC and may be constructed with conventional microprocessor components. The LIN will probably be a bus system, depending upon the exact microprocessor that is selected for the GDP and the organization that is cheaply available. The M is to be constructed with conventional microprocessor memory technology. The microprocessor will have a local cache memory. There may be more than one GDP in a PC; for example, the Intel system 432/600 [6] permits up to five GDPs to be connected to a single LIN.

Other devices to be connected to the LIN include I/O processors (IOP) and attached processors (AP) (via an IOP). More than one IOP or AP may be connected to the LIN to best meet the performance requirements of the PC. I/O processors perform two functions: communications with peripheral devices and communications with the GIN. They may contain buffer memories to deal with high speed or congested devices on interconnection networks.

The APs are special purpose processors which may be used to enhance

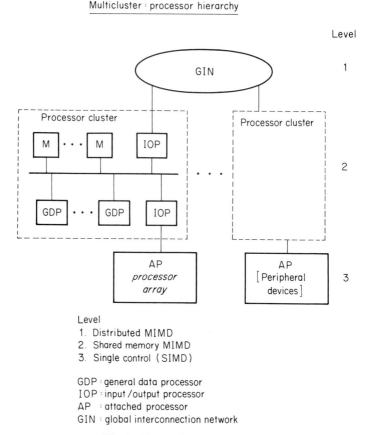

Fig. 1. Multicluster organization.

performance for specific applications. For example, as the floating-point performance of current VLSI microprocessors is still not very good, a fast floating-point processor may be a very useful addition to the PC. Other possibilities for APs are vector pipelines, systolic arrays and small processor arrays. There is no need for all PCs to have the same configuration; some PCs may be considered to be "experts" for certain applications by virtue of their APs.

2.2 Intel 432 Multicluster testbed

An experimental testbed using Intel 432 processors has been constructed for

Multicluster MIMD system

research into Multicluster systems. The Intel 432 is a 32-bit microprocessor designed for multiprocessor applications; it is programmed in Ada. The testbed consists of two clusters connected together by an 8086-based microprocessor system as shown in Fig. 2. Each cluster is controlled by an 8086-based microprocessor system. A cluster consists of an Intel 432/670 system. This is capable of supporting three general data processors (GDPs) and two I/O processors (IOPs) with a shared memory environment. With an extended backplane five GDPs can be supported. All processors have an equal priority to access the memory; access is controlled by a round robin scheduler. The data processing tasks on the system are shared by the GDPs.

Each IOP locally runs a conventional multitasking operating system on the 8086 and its own memory resource. Tasks running on the IOP have

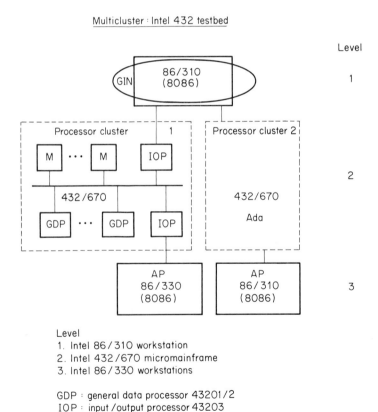

Fig. 2. Intel-432 based Multicluster testbed.

access to the 432/670 main memory and may communicate via interprocessor messages with tasks running on the 432 system. In the usual mode of operation the control systems are used to control the operation of their cluster and to record results from experiments. The communications system is used to transfer messages between the clusters and for data-file storage for both clusters.

The operation of the system is controlled by means of commands given at the three consoles of the 8086 systems. If the consoles are conventional terminals then the system is rather unwieldy to use, since an operator must coordinate all operations between these systems. Furthermore, the 8086 systems are heavily loaded and are unable to communicate with each other when managing the clusters. A single UNIX workstation is being installed to control all three console ports. The user now only requires one terminal, and command files can be used to control experiments and results can be easily recorded in a single file.

The operating system running on the 432/670, called iMAX, provides memory management, interprocessor message passing and task-scheduling capabilities. It is implemented as a set of Ada packages. Memory management includes garbage collection, which can run as a separate task concurrently with data-processing tasks, and memory compaction, which runs on demand.

Task scheduling is controlled by three parameters for each task: priority, deadline and time slice. Tasks that are ready to run wait on a dispatch queue until a processor becomes available. A processor becomes available when the task running on it terminates or its execution time exceeds its time-slice parameter. The priority parameter is used to determine which tasks can be run; the next task to be run is selected from all tasks with the highest priority. The deadline parameter is used to select between waiting tasks with the highest priority. Conceptually, the deadline is an indication of how long the task should have to wait for dispatching, relative to other tasks with the same priority. If two tasks arrive at the dispatching queue at approximately the same time, then the task with the smallest deadline will be scheduled to run first.

2.3 Intel 432 cluster performance

The Intel 432 system involves a 32-bit microprocessor designed for multiprocessor operation. The time for an elemental arithmetic operation is in the range of 1 to 25 μs. The cache size is small, therefore the cost of context switching is also small. In our programs the whole problem is stored in the local memory and the virtual memory scheme is not used.

It might be assumed that a task rendezvous (i.e. the calling of one task by

Multicluster MIMD system

another) would require more time than a procedure call, which is local to a task; this is not necessarily true. We measured the procedure call-and-return time for a simple procedure to be 0·6 ms. The 432 is an object-based system and procedure calls are dynamic, which means that each time a call is made a memory object must be allocated for the data declared within the procedure; this is the reason for the slow procedure call time. If a large amount of memory is required by the procedure then the call time may be significantly longer. Once a task is activated, however, no memory needs to be allocated for a task rendezvous; a request for rendezvous is simply enqueued on the entry queue of the task being called. An exact measure of the processor time for a task rendezvous is difficult to measure owing to task scheduling. A further consideration when using tasks should be given to the task creation time; we measured this to be 691 ms for the Intel 432. Since task creation

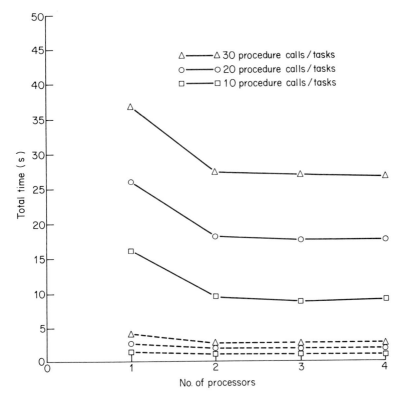

Fig. 3. Comparison of the total time between the procedure-based program (dashed lines) and the task-based program (solid lines) when the procedure/task execution time is 100 ms.

requires so much time, a tasking strategy that creates all needed tasks and reuses them as much as possible is to be preferred to a scheme that dynamically creates tasks when needed and disposes of them once a function has been performed.

A comparison between tasking and procedure calls for a simple task is made in Figs. 3 and 4. The broken lines indicate the execution time for a sequence of procedure calls and the solid lines are for the same work implemented by a number of tasks, each task performing the same amount of computation as a procedure call. In Fig. 3 the execution time for each task/procedure call is about 100 ms, in Fig. 4 the execution time is about 6 s. about 100 ms, in Fig. 4 the execution time is about 6 s.

A system task, the garbage collector, is always running on the system; This task is able to dynamically reclaim memory while other tasks are in progress. In Fig. 3 we see that the procedure call method is much faster than

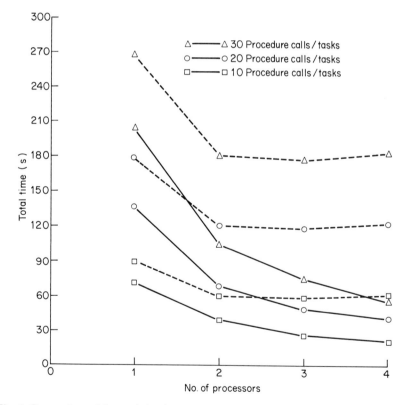

Fig. 4. Comparison of the total time between the procedure-based program (dashed lines) and the task-based program (solid lines) when the procedure/task execution time is 6 s.

Multicluster MIMD system

using tasks; this is because the task method is dominated by the task creation time. There is a significant speed improvement when two processors are used because the garbage collection is then running concurrently with the main problem. Additional processors have no effect on the procedure cases. For the task cases a slight improvement is noted with additional processors, but task creation, which is done serially, still dominates the computation time.

In Fig. 4, where the task/procedure computation time is much larger, we see that the task scheme is superior to the procedure scheme; the main reason for this is that the task creation time is now small compared with the task computation time. For the single-processor case tasking is faster because a large number of tasks are competing with the garbage collector, each with equal priority, while for the procedure scheme only one task competes with the garbage collector. When two processors are used with the procedure scheme the main program runs on one processor and the garbage collector runs on the other; there is no further improvement when more processors are added. For the tasking scheme there is an improvement with each additional processor, since there are always many tasks waiting to be run.

The tasking behaviour of the Intel 432 is further illustrated in Figs 5 to 8.

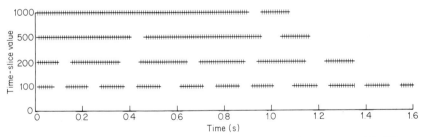

Fig. 5. Execution profile of the single-task program on a single processor with different time slice values.

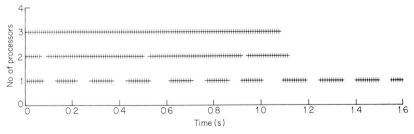

Fig. 6. Execution profile of the single-task program with different numbers of processors (time-slice value is fixed at 100 ms).

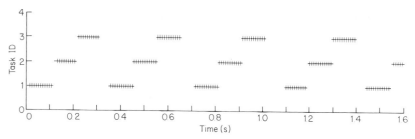

Fig. 7. Execution profile of the three-task program on a single processor.

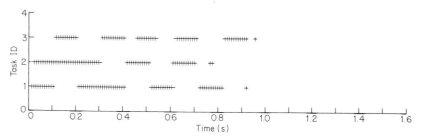

Fig. 8. Execution profile of the three-task program on two processors.

In these graphs the abscissa is real time and a + symbol indicates that a task is running and blank space indicates that the task is not running. In Fig. 5 the execution of a single task on a single processor for different time slice values is shown. In this case, the user task is competing directly with the garbage collector and other system tasks. In Fig. 6. the effect of using multiple processors for a single user task is shown. We see that even with two processors a small amount of time is lost to system tasks, but that the main task runs continuously with three processors. Figure 7 shows three user tasks running on a single processor; the garbage collector task is run between the end of task 3 and the start of task 1. The time slice for all user tasks is 100 ms. In Fig. 8 the effect of adding a second processor to the three-task problem is shown.

3 NEAR-NEIGHBOUR ALGORITHMS

Multicluster tasking strategies for near-neighbour algorithms have been considered in detail. Near-neighbour algorithms are used in a wide variety of applications including low-level image processing. They have attractive properties for this study in that they are deterministic, they exhibit locality

Multicluster MIMD system

of computation, and they can be partitioned into subtasks in a large number of different ways.

A near-neighbour algorithm is usually expressed as a long sequence of near-neighbour operations. A near-neighbour operation generates a result matrix having the same dimensions as the input matrix (or matrices). Each element of the result matrix is computed from the corresponding element in the input matrix and the immediately adjacent near-neighbour elements.

3.1 Near-neighbour algorithms on Multicluster

The distribution of near-neighbour algorithms on a distributed Multicluster environment has been considered in [2] and [3] for mesh, multistage and other interconnection schemes. The general task allocation strategy is to partition the matrix into square submatrices and allocate each submatrix to a cluster. For each near-neighbour operation the edges of the submatrices are transferred between clusters.

A very important consideration is fault tolerance; the tasking scheme was modified to dynamically reconfigure the matrix partitions to distribute the

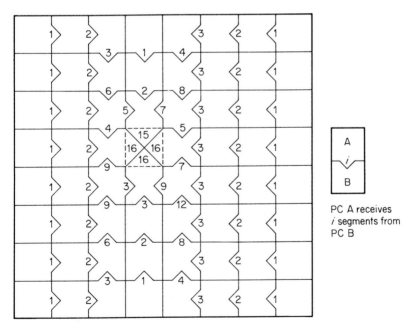

Fig. 9. Data partitioning of the Multicluster system with a single faulty PC (dashed lines indicate the faulty PC data submatrix).

load evenly on the remaining clusters after a cluster fault. This reconfiguration strategy is illustrated in Fig. 9 for a system of 64 PCs in which a single faulty PC is at location (4,4). The load associated with this PC must be redistributed amongst the remaining 63 PCs. That is, each PC must receive one segment of additional data elements, which is 1/63 of the data elements allocated to the faulty PC. With the redistribution of data shown in Fig. 9 the number of interprocessor communications is minimized and the load is balanced for each cluster. Further reconfiguration is possible if other PCs become faulty.

As a result of this reconfiguration the regions to be processed by each cluster are no longer square. The above scheme considers each cluster as a lumped processing resource. It could be easily modified to deal with faulty processors within clusters such that the load assigned to each cluster is proportional to the number and speed of the working processors it contains.

3.2 Near-neighbour algorithms on a single cluster

First we consider the case when the entire task is to be processed on a single cluster; in this case there is no intercluster communication. The goal is to minimize the execution time for a near-neighbour operation. The main parameters in this case are the number of tasks, the task time slice and task priority. More formally, the problem is to compute a result matrix of M elements from an input matrix of the same dimensions. The image function may be considered as a sequence of matrix operations, each operation consisting of M independent identical elemental operations. On a multiprocessor system with N processors the problem must be split into a number of K subtasks. To efficiently implement this problem on a cluster the optimal value for K and a scheduling strategy are needed.

The minimum value for K is N. In this case one task will be allocated to each processor. The main problem with this scheme is that, unless time slicing is used, the fault tolerance is not good; for example, if a single processor fails then the time to completion will be doubled. No allowance has been made for system tasks such as the garbage collector, which may cause an imbalance in processing when time slicing is not used.

A second extreme possibility is to create M tasks, one for each element. The problem with this scheme is that the overhead required to create and manage the very large number of tasks on a conventional processor will far exceed the data computation time. This is a similar approach to the fine-grain data flow scheme; typically with such systems each data processor is augmented with a second special processor for task (operation) management.

Multicluster MIMD system

Finally we consider the possibility of K being significantly greater than N but still much smaller than M. If K is much larger than N then time slicing will not be needed to balance the load and a considerable saving in overhead due to context switching is possible; however, if K is not much larger than N time slicing will still be necessary. It is difficult to predict, *a priori*, what will be the optimal number of tasks. Empirical results for the Intel 432 system are shown in Fig. 10. The task creation time is not included in this figure since this may be done once during system initialization. For these results garbage collection is running and all user tasks have a time slice of 100 ms. For this system and this application the best results are obtained by splitting the problem into 10 to 15 tasks.

3.3 Task decomposition for multiple clusters

For multiple-cluster operation each cluster receives edge information from adjacent clusters. Two types of tasks can be identified: tasks that only use

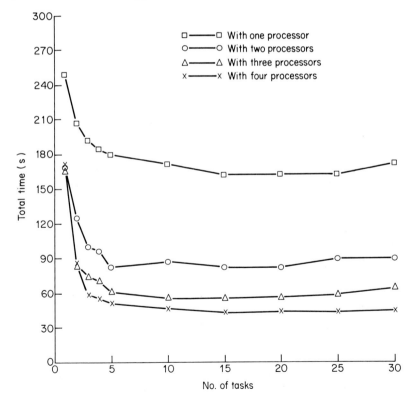

Fig. 10. Total time as a function of the number of tasks in a program.

local information and tasks that require external edge information. Since the latter tasks will start at a later time they can be given a higher priority or a longer time slice in order to balance the load in the cluster.

4 CONCLUSION

Simple tasking strategies for a Multicluster system have been presented. An effective task-decomposition strategy for an MIMD system for even simple computer-vision algorithms is very complex and depends upon a number of system parameters. However, given the *a priori* information that is available with many problems, it should be possible to devise a set of effective strategies.

It should not be necessary for the user to program the task decomposition, which is the current practice with SIMD systems. For example, for a near-neighbour problem a user should only have to specify the near-neighbour operations. The allocation of data elements to processors should be done by the system. Much more work is needed to develop effective task-decomposition strategies. The availability of such strategies will be the key to the use of MIMD systems for computer-vision applications.

REFERENCES

[1] Reeves, A. P. (1985). Multicluster: an MIMD system for computer vision. In *Integrated Technology for Parallel Image Processing* (ed. S. Levialdi), pp. 39–56. Academic Press, London.
[2] Uyar, M. U. and Reeves, A. P. (1985). Fault reconfiguration for the near neighbor problem in a distributed MIMD environment. In *Proc. 5th Conf. Distributed Computing*, pp. 372–379.
[3] Uyar, M. U. and Reeves, A. P. (1985). Fault reconfiguration in a distributed MIMD environment with a multistage network. In *Proc. 1985 Int. Conf. on Parallel Processing*, pp. 798–806.
[4] Reeves, A. P. (1983). Fault tolerance in highly parallel mesh connected processors. In *Computing Structures for Image Processing* (ed. M. J. B. Duff). pp. 77–94. Academic Press, London.
[5] Batcher, K. E. (1980). Design of a massively parallel processor. *IEEE Trans. Comp.* **29,** 836–840.
[6] Intel Corporation (1981). *Intel System 432/600 System Reference Manual*, Order no. 172098-001.

Chapter Thirteen
Multifunction Processing with PASM

J. T. Kuehn and H. J. Siegel

1 INTRODUCTION

Computer architects have proposed and developed a wide variety of processor interconnection strategies, memory access schemes, and process/processor synchronization mechanisms for parallel processing. The resulting machine structures are largely application-driven. For example, parallel computers with two-dimensional arrays of processors, memories local to the processors, and nearest-neighbour interprocessor communication (Fig. 1a) have been successfully applied to image processing applications (e.g. MPP [1], CLIP4 [2]). In this case, the processor arrangement and interconnection topology match the processing requirements of the raw image pixel data. For image-understanding applications, pyramids of processors (Fig. 1b) have been proposed [3,4]. Processors in the base of the pyramid operate on the lowest level of image data abstraction (pixels), those at middle level(s) in the pyramid operate at higher data abstractions (lines, contours, shape and texture), and at the top level the highest-level functions of recognition are performed (e.g. identifying an enemy tank moving north at 12 kilometers per hour). Here again, the machine topology matches the character of the data and the flow of the task up the pyramid. For execution of functional languages, processors arranged in binary trees (Fig. 1c) have been used (e.g. DADO [5]). The tree naturally expresses the recursive subdivision of goals into subgoals.

The application drives the design of the processor/memory interface as well. Private processor memories such as those found in MPP are appropriate when memory references are highly localized. As an example, an image-smoothing algorithm that transforms a pixel and its eight nearest neighbour pixels to a single smoothed output pixel exhibits a (spatially)

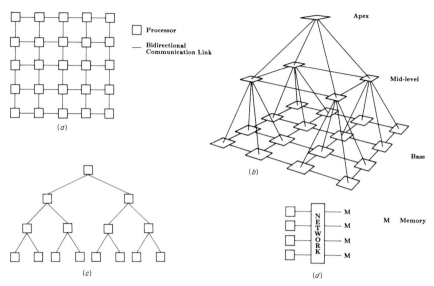

Fig. 1. A variety of parallel computer organizations: (a) mesh-connected; (b) pyramid; (c) full binary tree; (d) shared-memory system.

localized memory reference behaviour. Shared memory systems, in which the processors can access any memory, are more effective when the access patterns are either random or non-localized. In such systems (e.g. NYU Ultracomputer [6]), a bidirectional interconnection network connects the processors on one side of the network to the memories on the other side (Fig. 1d).

The mode of parallel computation chosen for an algorithm is also affected by the character of the data. Two common parallel modes are the single-instruction-stream–multiple-data-stream (SIMD) mode and the multiple-instruction-stream–multiple-data-stream (MIMD) mode [7]. In SIMD machines there is typically a control unit (CU) that operates on scalar data, a collection of processors with private or shared memories, and an interconnection network. The CU broadcasts instructions to the processors, and all enabled processors execute the same instruction at the same time, but each on its own data. Broadcast instructions form the single instruction stream; processors operate on the multiple data streams. Examples of SIMD machines are the Illiac IV [8] STARAN [9], CLIP4 [2], and MPP [1]. Image processing was a key application cited for each of these machines. MIMD machines are collections of processors with private or shared memories and an interconnection network. The processors of an MIMD machine operate independently and asynchronously. Examples of MIMD machines are C.mmp [10] Cm* [11], NYU Ultracomputer [6], Butterfly [12], and RP-3 [13]. ZMOB [14] is an MIMD machine constructed for image-processing applications.

SIMD parallelism has shown tremendous performance for structured tasks with large data sets. The early stages of image-processing tasks that consider the pixel-level representation of an image such as clipping, smoothing, histogramming and two-dimensional correlation are examples of highly efficient SIMD algorithms. MIMD parallelism is more suited for the later stages of image-processing tasks that deal with higher-level image representations such as lines, contours and regions. Some machines are being designed to work in both modes of parallelism so that the same processors can be used to perform complete tasks. Also, each of the algorithms that comprise the task may be coded to utilize the most efficient mode of parallelism. Examples of machines that can support both the SIMD and MIMD modes of parallelism are TRAC [15], PASM [16] and DADO [5]. Some are partitionable, allowing multiple independent SIMD machines to operate concurrently (e.g. University of Washington Pyramid [17]), or supporting multiple independent SIMD and MIMD machines (e.g. TRAC, PASM, DADO).

It is clear that to support the computational needs of the varied tasks in an application domain such as image understanding, a *multifunction* computer is desirable. The "ultimate" multifunction computer would provide to users a virtual machine that can be configured with any number and arrangement of virtual processors and memories and a variety of memory access capabilities, interconnection structures, processing modes and synchronization mechanisms. Of course, a sophisticated simulator running on a conventional sequential machine could provide the requisite virtual machine; however, there is a substantial performance limitation for this approach. Our goal is the design of a flexible multifunction parallel-processing system, PASM, that can be dynamically reconfigured to meet the particular processing needs of a large variety of applications in the image-understanding domain.

PASM (*partitionable SIMD/MIMD*) [16] is a multifunction research parallel-processing system being developed at Purdue University. It can be configured under user software control to operate in either SIMD or MIMD mode and to switch between modes during execution of a program. PASM's multistage interconnection network provides for the establishment of flexible interconnection patterns among the processors of the system. This allows emulation of a variety of machine topologies: meshes, trees, pyramids, rings, and so on. The arrangement of processors and the nature of the interconnection network allow PASM to be partitioned into one or more virtual machines of various sizes, each of which operates independently of the others. Although PASM processors have local memories and use a message-based communication scheme, PASM can emulate a shared memory system. How the PASM hardware and operating system support the machine's reconfiguration is the subject of this chapter.

Section 2 overviews the PASM architecture and the implementation of a 30-processor PASM prototype. PASM's multifunction interconnection network is described in Section 3. The mechanism that PASM uses to dynamically switch from SIMD to MIMD mode is discussed in Section 4. How PASM emulates shared memory is considered in Section 5.

2 PASM OVERVIEW

2.1 General design concepts

A block diagram showing the basic components of PASM [16] is given in Fig. 2. The *system control unit* is a conventional computer and is responsible for the overall coordination of the activities of the other components of PASM. The *parallel computation unit* contains N processing elements (PEs) and a multistage interconnection network. Each PE consists of a processor and a private memory module. PEs are numbered from 0 to $N-1$ and each PE knows its number (address). The *memory management system* controls the loading and unloading of the PE memory modules from the multiple secondary storage devices of the *memory storage system*. The *micro controllers* (MCs) are a set of Q microprocessors which act as the control units for the PEs in SIMD mode and orchestrate the activities of the PEs in MIMD

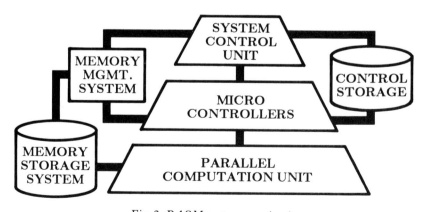

Fig. 2. PASM system organization.

mode. Each MC controls N/Q PEs. *Control storage* contains the programs for the MCs.

PASM is being designed to have $N = 1024$ PEs and $Q = 32$ MCs. The PASM prototype, illustrated in Fig. 3, is being constructed with $N = 16$ PEs and $Q = 4$ MCs.

Partitions

The MCs are the multiple control units needed in order to have a partitionable SIMD/MIMD system. There are $Q = 2^q$ MCs, physically addressed (numbered) from 0 to $Q - 1$. The partitioning rule for PASM requires that all PEs in a partition agree in the low-order bit positions of

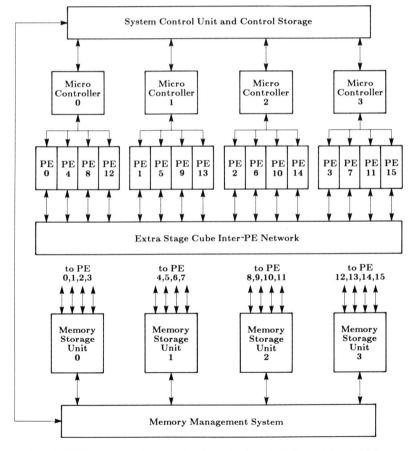

Fig. 3. *PASM prototype organization with $N = 16$ PEs and $Q = 4$ MCs.*

their addresses. In particular, the physical addresses of the N/Q processors that are connected to an MC must all have the same low-order q bits. The value of these low-order q bits is the physical address of the MC. An SIMD machine partition of MN/Q processors, where $M = 2^m$ and $0 \le m \le q$, is obtained by having M MCs use the same instructions and synchronizing the MCs. The physical addresses of these MCs must have the same low-order $q - m$ bits so that all of the PEs in the partition have the same low-order $q - m$ physical address bits. Similarly, an MIMD machine partition of size MN/Q is obtained by combining the efforts of the PEs associated with M MCs that have the same low-order $q - m$ physical address bits. Q is the maximum number of partitions allowable, and N/Q is the size of the smallest partition.

In general, the possible advantages of a system where partitions of various sizes can be formed include the following.

(i) *Variable machine size for efficiency.* If each MC controls N/Q PEs, and Q MCs are available, partitions of N/Q, $2N/Q$, ..., N PEs can be formed by coordinating the efforts of more than one MC.

(ii) *Fault detection.* For situations where high reliability is needed, three partitions can run the same program on the same data and "vote" on the results.

(iii) *Fault tolerance.* If a single PE fails, only those partitions that must include the failed PE need to be disabled in SIMD mode. The partitions including the failed PE(s) may continue to be used for MIMD mode programs (even with multiple PE failures), but degraded performance results for programs that would normally utilize more than the number of still-functioning PEs.

(iv) *Multiple simultaneous users.* Since there can be multiple independent partitions, there can be multiple simultaneous users of the system, each executing a different SIMD and/or MIMD program.

(v) *Program development.* Rather than trying to debug a parallel program on, for example, 1024 PEs, it can be debugged on a smaller size partition of 32 or 64 PEs.

(vi) *Subtask parallelism.* Two independent subtasks that are part of the same job can be executed in parallel, sharing results if necessary.

Secondary memory organization

The *memory storage system* provides secondary storage space for the PE data files in SIMD mode and for the PE data and program files in MIMD mode.

It consists of N/Q independent *memory storage units* (MSUs), numbered from 0 to $(N/Q) - 1$. Each MSU consists of a high-capacity disk drive, disk controller, and a microprocessor to manage the file directory system on the disk. Each MSU is connected to Q PE memory modules. For $0 \le i < N/Q$, MSU i is connected to those PE memory modules whose physical addresses have the value i in their $n - q$ high-order bits. Recall that, for $0 \le k < Q$, $MC\,k$ is connected to those PEs whose physical addresses have the value k in their q low-order bits. This is shown for $N = 16$ and $Q = 4$ in Fig. 3.

The loading of data into a partition of MN/Q PEs requires only M parallel block loads if the data for the PE memory module whose high-order $n - q$ logical address bits equal i is loaded into MSU i. This is true no matter which group of M MCs (which agree in their low-order $q - m$ physical address bits) is chosen [18].

Interconnection network

The interconnection network is an N-input N-output switch that is controlled by the PEs in a distributed fashion. PE i, $0 \le i < N$, is connected to input port i and output port i of the unidirectional network. PASM will use a type of Generalized Cube network. The Generalized Cube network topology has $n = \log_2 N$ stages of $N/2$ interchange boxes [19]. Each interchange box is a two-input, two-output device which can be set individually to one of the four legitimate states shown in Fig. 4.

The connections in this network are based on the cube interconnection functions. Let $P = p_{n-1} \ldots p_1 p_0$ be the binary representation of an arbitrary I/O line label on an interchange box (Fig. 4). Then the n cube interconnection functions can be defined as

$$\text{cube}_i(p_{n-1} \ldots p_1 p_0) = p_{n-1} \ldots p_{i+1} \overline{p_i} p_{i-1} \ldots p_1 p_0,$$

where $0 \le i < n$, $0 \le P < N$, and $\overline{p}_i$ denotes the complement of p_i. This means that the cube$_i$ interconnection function connects P to cube$_i$ (P), where cube$_i$ (P) is the I/O line whose label differs from P in just the ith bit position. Stage i of the Cube topology contains the cube$_i$ interconnection function, i.e. it pairs I/O lines that differ only in the ith bit position.

A network interchange box is controlled by a routing tag which is the destination port address D [20]. Let $d_{n-1} \ldots d_1 d_0$ be the binary representation of D. An interchange box at stage i need only examine d_i. If $d_i = 0$ a connection is made from the interchange box input to the upper output link; otherwise, a connection is made to the lower output link. Thus, if the tag bits associated with a given interchange box are 0 on the upper input link and 1 on the lower input link, the box is set to the straight state. Similarly, if the tag bits are 1 on the upper input link and 0 on the lower input link, the box is set

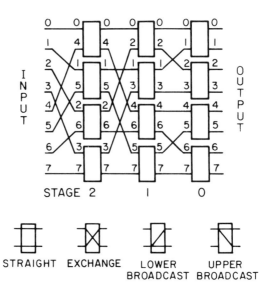

Fig. 4. Generalized cube topology shown for $N=8$ and the four legitimate states of an interchange box.

to the exchange state. Other combinations, e.g. 0 tag bit on both the upper and lower input, create "conflicts" in the network since no configuration of the box can make the desired connection.

Tags that can be used for broadcasting data are an extension of this scheme. An n-bit broadcast tag (B) indicates in what stages the boxes are to be placed in a broadcasting mode. For example, if bit b_i is a 1, a broadcast is performed; otherwise, the straight or exchange function specified by the normal tag is used.

2.2 Prototype design

A prototype of the PASM system is currently being constructed (see Fig. 3). This subsection contains information about the prototype hardware related to PASM's multifunction capabilities. Additional information about PASM and the prototype can be found in [18,21,22].

2.2.1 PE overview

A prototype PE consists of five VME-standard boards: a Motorola MC68010-based CPU board, two dynamic memory boards with a com-

Multifunction processing with PASM

bined capacity of 2 Mbytes, an interconnection network interface board, and an I/O board for communication between the PE and its MC. The memory is divided into pages, each of which is mapped and protected by read/write/execute access bits. This allows sharing and secure coexistence of user and system routines. The interconnection network interface board contains parallel ports that are accessed by the CPU directly using conventional load/store instructions or by using a direct memory access (DMA) controller.

2.2.2 Interconnection network

The PASM prototype's interconnection network is a circuit-switched implementation of the Extra Stage Cube network [23], which is a fault-tolerant version of the multistage Cube network (see Fig. 5). This network is single-fault tolerant and has been shown to be very robust under multiple faults [24]. For the 16-PE prototype the Extra Stage Cube network consists of five stages of interchange boxes with eight boxes per stage. Circuit switching was chosen for its ease of implementation as well as its particular suitability (when compared with packet switching) for the anticipated large "conflict-free" data transfers under DMA control. Using a circuit-switched network, prior to any message transmission between a network source–destination pair, a physical path through the network must be made connecting the pair. The path is established through the use of a request-grant protocol. This connects the source–destination pair for the duration of the message transmission. Individual words within the message are transferred from the source PE to the destination PE using a handshaking protocol between the parallel ports that interface the PEs to the network.

2.2.3 MC overview

A prototype MC consists of seven VME-standard boards: a Motorola MC68010-based CPU board, two dynamic memory boards with a combined capacity of 2 Mbytes, an I/O board for communication between the MC and its PEs, a fetch-unit board, and two I/O boards for communication with other system components. The *fetch unit* is a finite state machine that operates only when an SIMD program is being executed. It is controlled by the MC CPU using commands directing it to fetch and broadcast instructions to the PEs. An MC's dynamic memory is accessible only to the MC CPU. It contains the scalar and control flow instructions for SIMD programs and the instructions to control the fetch unit for these programs. The dynamic memory is also used by the MC CPU to execute coordinating programs when the PEs are in MIMD mode. The fetch-unit memory is

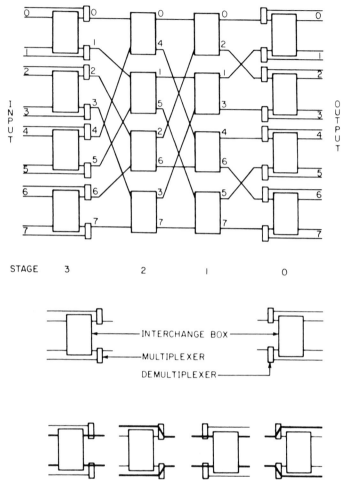

Fig. 5. *Extra stage cube network topology shown for N = 8.*

normally accessed by the fetch unit and contains only the instructions to be broadcast to the PEs in SIMD mode. The *PE instruction queue* contains a stream of instruction words that have been recently fetched and enqueued by the fetch unit and are ready to be broadcast to the PEs. When all of the PEs in the partition are ready for the next instruction, an instruction is dequeued and broadcast to them. The mechanism for requesting and broadcasting instructions in SIMD mode is indicated in Section 4.

3 PASM MULTIFUNCTION NETWORK

Prototype PEs communicate via the interconnection network using a sequence of I/O port read and write operations. A source PE S specifies where its data is to be routed by computing the routing and broadcast tags discussed in the last section. The two tags and the source PE number (S) are written to a parallel I/O port which instructs the network to set switches to make a connection with the destination address(es). The source tag is not used in setting switches, but is needed by the destination PE in MIMD mode to determine where the incoming data originated.

Having set up the path, a source PE can now send messages to the destination PE(s). Data transmissions pass through two parallel I/O ports called data transfer registers (DTRs) [16]. The two ports are configured as 16-bit unidirectional channels. One direction connects the PE to the network input (DTRin), and the other to the network output (DTRout). The data to be transmitted is written to the DTRin port, signalling to the network that the transfer should be made. Subsequent transfers route items to the same destination until the "network setting" is changed. Incoming data from other PEs arrives on a PE's DTRout port. In SIMD mode all enabled PEs do the setting, transmitting and receiving operations at the same time. In MIMD mode PEs use the network asynchronously. Therefore in MIMD mode incoming messages cause the receiving CPU to be interrupted so that it can retrieve, interpret and handle the message.

The *partitionability* of a network is its ability to divide the system into independent subsystems of different sizes [19]. The Cube network can be partitioned into independent subnetworks of various sizes where each subnetwork of size $N' \leq N$ will have all of the connection properties of a Cube network built to be of size N'. In PASM the partitioning is accomplished by requiring that the addresses of all of the network I/O ports in a partition of size 2^i agree (have the same values) in their low-order $n - i$ bit positions. For the example in Fig. 6, subnetwork A consists of ports 0, 2, 4 and 6; subnetwork B consists of ports 1, 3, 5 and 7. All ports in subnetwork A have a 0 in the low-order bit position; all ports in subnetwork B have a 1 in the low-order bit position. By setting all of the interchange boxes in stage 0 to straight, the two subnetworks are isolated. This is because stage 0 is the only stage that allows input ports that differ in their low-order bit position to exchange data. Each subnetwork can be separately further subdivided resulting in subnetworks of various sizes. This network partitioning property allows the set of PASM PEs to be divided into independent machine partitions of various sizes. It also isolates potentially careless or malicious users running on different partitions from each other.

The flexibility of multistage networks such as the Generalized Cube is

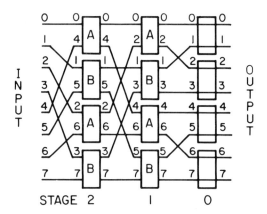

Fig. 6. *Cube network of size eight partitioned into two subnetworks of size four based on the low-order bit position.*

primarily due to the characteristic that any one processor can communicate with any other single processor in its partition in only one set-transfer step (assuming no conflicts with other paths in the network). Also, a large variety of useful network permutations (simultaneous transfers of N data items from input to output ports) can be performed. This is in contrast with networks such as the ones shown in Fig. 1(a–c) in which communication between two non-adjacent processors may require many steps. On the other hand, the fixed interconnections illustrated in Fig. 1(a–c) have an advantage in that they require no "setup" overhead for path establishment since the links are always "there". Also there is only one outgoing link per processor when a multistage network is used; there may be as many as four outgoing links per processor for the mesh-connected structure of Fig. 1(a) and up to $b + 5$ outgoing links per processor in the mesh-connected middle-levels of a pyramid with a fanout of b. Therefore, while the multistage network can emulate any single outgoing link, it cannot emulate all of them simultaneously. To get the effect of using a different outgoing link with a multistage network, the processor must request a change in the network setting. This flexibility-speed tradeoff indicates that PASM will emulate the interconnection structures illustrated in Fig. 1(a–c) with varying efficiency.

To demonstrate some of the flexibility of the PASM multistage network, consider its use in emulating four interconnection structures: a ring, a mesh, a pyramid and a tree. Aspects to be considered are the ease with which a PE can calculate the addresses of its neighbours and the machine sizes that make the mapping of PASM PEs onto the emulated structure efficient. For the

Multifunction processing with PASM

following examples it is assumed that the physical number of PEs available equals or exceeds the size of the virtual machine to be emulated. If the virtual machine size exceeded the physical machine size some PEs would need to emulate more than one virtual PE, necessitating multiple network transfers for a given data movement.

In a ring-connected structure of size P, PE i is connected to PE $i-1 \mod P$ and PE $i+1 \mod P$. The destination tags needed to set the interconnection network are easily calculated in this case. If P is a power of two the network setting is especially trivial and free from conflicts as long as all PEs are talking to their "$+1$" neighbour or their "-1" neighbour simultaneously. PEs may also communicate with their neighbours in even–odd pairs if P is a power of two. If P is not a power of two a simultaneous transfer "to the right" or "to the left" is not possible owing to conflicts in the network. Such communication would require two network set-transfer steps: one for PE i to PE $i+1$ ($0 \le i < P-1$) and one for PE $P-1$ to PE 0 (assuming the "to the right" transfer).

For the mesh-connected structure of R rows and C columns of processors, PE i is connected to PE $i - C$ to the north, PE $i+1$ to the east, PE $i+C$ to the south and PE $i-1$ to the west. (Wraparound connections such as those permitted by MPP [1] are ignored.) Here again, the addresses of a PE's neighbours are easy to calculate. Also, any simultaneous transfer, be it north, east, south or west, is achievable in one step (ignoring wraparound) regardless of R and C. Wraparound connections are possible without conflict, but only if R and C are both powers of two.

Many types of communication patterns occur in the pyramid structure. Within a level, PEs can communicate with their mesh-connected neighbours, PEs in a higher level can broadcast to the b PEs below them (where b is the fanout), and, finally, communication between a parent and one of its b children is possible. These varied patterns make mapping the pyramid onto the set of PASM PEs more difficult. The first constraint is that the PASM network can broadcast only to a number of PEs that is a power of two; therefore b must be a power of two for the pyramid structure to be efficiently emulated. One solution is to use $2(b^{L-1})$ PASM PEs for an L-level pyramid, where $L > 1$ and b is an even power of two. An example of a conflict-free numbering for the PEs configured as a 3-level pyramid with $b = 4$ is shown in Fig. 7.

The tree configuration is a special case of the pyramid. The fanout of b is attainable without conflict if and only if b is a power of two [25]. However, since the intra-level mesh connections are not present in the tree, the addressing is simplified greatly. One natural ordering is to assign the PE 1 to be the base, PEs 2 and 3 its children, PEs 4 and 5 the children of PE 2, PEs 6 and 7 the children of PE 3, and so on.

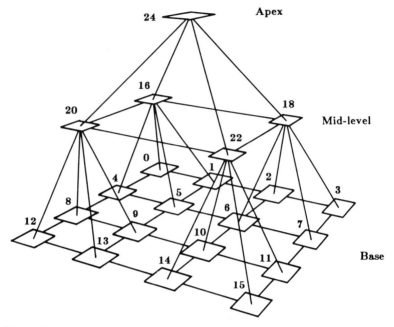

Fig. 7. *Numbering scheme for a three-level pyramid structure with fanout of four.*

4 MULTIFUNCTION OPERATION OF PEs

A partition may act either as an SIMD or as an MIMD machine. This section describes how the off-the-shelf CPUs in the PASM prototype behave in both modes and how they dynamically switch between modes.

4.1 MIMD

In MIMD mode the PE operates as a conventional machine, fetching both instructions and data from its own memory. The MC68010 PE CPU used in the prototype generates 24-bit addresses which are decoded by hardware as references to private memory, EPROM, I/O devices, and so on, just as in a conventional single-CPU microcomputer. Each address is accompanied by a read/write indicator and function codes indicating whether the reference is to "data" or "program" area and if the CPU is operating in "user" or "supervisor" (more privileged) state. When an address is placed onto the address bus and the bus cycle is a "read", address-decoding logic determines which PE local memory or I/O device is being referenced, that

Multifunction processing with PASM 223

device is enabled, and the data requested is placed onto the data bus by the device. The MC68000-family CPUs use an asynchronous bus cycle; $\overline{DTACK}$ is the data transfer acknowledge signal generated by the device being referenced that indicates to the MC68010 CPU that the requested data is on the bus and that the bus cycle can be terminated by latching in the data. The item on the data bus is an instruction and is latched into the CPU instruction register if a program reference was being performed; otherwise, data is obtained. For "write" cycles $\overline{DTACK}$ indicates that the data to be written has been latched into the device.

4.2 SIMD

A certain range of addresses (the *SIMD instruction space*), when accessed with a program reference, indicates that the next instruction is to come from the MC fetch unit rather than from the local PE memory. Therefore the only difference between a PE operating in SIMD and MIMD modes is that in SIMD mode the CPU program counter contains an address in the SIMD instruction space, while in MIMD mode the program counter points elsewhere.

As discussed above, when a PE accesses the SIMD instruction space a request for an SIMD instruction is made to its controlling MC. The set of N/Q SIMD instruction request signals (one from each PE associated with a single MC) are and-ed together to determine when all PEs in the group are ready to receive the next instruction. SIMD instruction request signals from inactive PEs are forced to be "requesting" to support this protocol. When multiple MCs are being used to control the PEs in a partition, the SIMD instruction request signals of each participating MC are further and-ed together to determine when all PEs in the complete partition are ready to receive the next instruction. When all active PEs in a partition request an SIMD instruction, an instruction word is broadcast from the MC(s) of the partition, placing it on all of the PE data busses simultaneously. Together with the instruction broadcast, a $\overline{DTACK}$ signal is generated and broadcast by the MC. Each enabled PE latches the instruction into its local instruction register, decodes it and performs the operation or requests additional operand words. Inactive PEs are prevented from "seeing" the $\overline{DTACK}$ signal provided by the MC; therefore they do not know that the bus cycle is over and they never begin to process the instruction provided. If the address-decoding logic determines that a PE local memory or I/O device address is being referenced, no SIMD instruction request is generated and the $\overline{DTACK}$ signal is provided by the device being referenced just as in a conventional machine.

Since the execution time of some machine instructions is data-dependent (e.g. integer division), some PEs might complete an instruction before others. The "faster" PEs would request the next instruction by entering a bus read cycle and would have to wait a number of clock cycles before the other PEs completed their operations. For this reason, the standard bus watchdog timer used to detect references to unimplemented memory is disabled during SIMD program accesses.

In SIMD mode, the internal PE CPU program counter serves only to identify a request for an instruction word. The actual value of the PE program counter is irrelevant, as long as it references the SIMD instruction space. However, the program counter is incremented automatically upon receiving an instruction word, and will eventually near the end of the SIMD instruction space. Therefore the PE program counter needs to be periodically reset to the beginning of the space; otherwise, the PE program counter would "fall off the end" of the SIMD instruction space and try to fetch instructions locally. The PE program counters are reset periodically by the MC by enqueuing a "jump" instruction to the PE instruction queue. The mechanism involved in the operation is discussed in [26]. Since the SIMD instruction space is large, overhead in resetting the PE program counters is minimal.

4.3 Mode switching

Earlier, it was indicated that some of the constituent algorithms of an image-processing/understanding task may be more efficiently coded to use the SIMD mode of parallelism, while others are expressed more naturally as MIMD algorithms. Therefore it is advantageous for the machine to be able to switch modes between the algorithms in a task. Examples shown in this subsection will also demonstrate that it may be advantageous for the machine to switch modes during execution of a single algorithm.

To effect a change from SIMD to MIMD mode the MC(s) controlling the partition's PEs broadcast a "jump" instruction to some address within the PE local memory space. Typically, this address would be the beginning of a program stored in ROM that would initialize the PE operating system kernel for MIMD processing. While in MIMD mode PEs do not access the SIMD instruction space since MIMD instructions and data are contained wholly within a PE's memory. When the PE is ready to revert to SIMD mode it jumps to the beginning of the PE instruction queue space. When all of the PEs have done this, SIMD processing continues because all of the PEs have indicated their readiness for an SIMD instruction.

Using the same mechanism, the MC can instruct PEs to temporarily enter MIMD mode during the execution of an SIMD program. This might be done for a calculation for which the execution time is highly data-

Multifunction processing with PASM 225

dependent, for example, converging on a solution by successive iteration. In this case, the MC would broadcast a "branch to subroutine" instruction to the PEs. The subroutine to perform the MIMD calculations would be in the PEs' memories and would be executed by each PE independently as soon as the "branch to subroutine" instruction was encountered. The subroutine would be ended by a "return" instruction, at which point the PEs would "rendezvous" and continue the SIMD program.

Another example of how switching modes during the course of a program may increase efficiency is illustrated by the following. Consider the parallel equivalent of an "if-then-else" construct:

　　　　if (< parallel-expression >)
　　　　　　< then-block >
　　　　else
　　　　　　< else-block >

the < parallel-expression > is an expression that depends on a variable that may have a different value in each PE. Therefore, the expression will be true in some PEs and false in others. Those PEs that found the expression to be true should execute the < then-block >, while the rest should execute the < else-block >. In a strict SIMD environment the < then-block > and the < else-block > cannot be executed by the PEs simultaneously since there is a single instruction stream. Thus side effects from < then-block > may affect the execution of < else-block >. In this case, PASM can evaluate the expression in SIMD mode and then temporarily switch to MIMD mode so that the < then-block > could be executed in parallel with < else-block >. The PEs would rendezvous after executing the < then-block > or < else-block > and continue processing in SIMD mode. By handling conditionals in this way, the execution time of the conditional is the maximum of the < then-block > time and the < else-block > time, rather than their sum.

The MC can also enforce synchronization of MIMD programs using the SIMD instruction space. Suppose that, at the synchronization point in the MIMD program, PEs executed a "branch to subroutine" instruction where the branch destination is an address in the SIMD instruction space. The MC meanwhile sends a "return" instruction to the queue. As each PE encounters the synchronization point and attempts to fetch an SIMD instruction, it is forced to wait until all PEs have reached the point. When the last PE arrives at the synchronization point the "return" instruction is released and the PEs continue their MIMD programs.

5 MESSAGE PASSING AND SHARED MEMORY

Private (local) processor memories have typically been employed in parallel processors being designed for image-processing applications (e.g. MPP, CLIP4, ZMOB). This is due to the large percentage of local data references as compared with global references. On the other hand, there are applications and algorithms for which it is desirable for one PE to have access to data in other PEs' memories. An example is an image-understanding algorithm that uses "global" knowledge to recognize certain patterns.

In SIMD mode such inter-PE communications are inherently structured and synchronized by the single instruction stream program. Consider the communication of data in PASM MIMD mode. To obtain a data item from a remote PE a transaction request must be generated, encapsulated in a message and sent to the remote PE. When the remote PE replies, the returned message is decoded and handled. In the PASM prototype, a DMA channel can be set up between two PE memories using the interconnection network as the communication medium. While this makes block data transfers between PEs more efficient since the PE CPU is not interrupted for each data item transferred, there is still some overhead involved in initializing the DMA controllers on each end of the transfer and in establishing the network path.

The message generation, transmission and handling can be done explicitly by the user or can be done by the operating system using the shared memory emulation technique described below.

Several recent MIMD machine designs have included both local and shared memory access capabilities. In the BBN Butterfly machine 24-bit addresses generated by the MC68000 CPUs are monitored by a microprogrammable bit-slice coprocessor. Some addresses are determined to be local and are handled by local dynamic memory or I/O ports; others are interpreted as global. Part of each Butterfly PE's memory is shared and appears as one contiguous address space within the aggregate of processors. Therefore seven of the higher-order bits of the global address identify the remote PE number (0 to 127), while lower-order bits indicate the shared memory word in that PE. When global addresses are generated a message indicating the address and the transaction to be performed is sent through the network to the remote PE holding the shared address. The coprocessor in the remote PE steals memory cycles to perform the transaction desired. If the transaction was a read the data is returned to the initiating PE's coprocessor (which remembers that a transaction was pending), which in turn provides the data to the CPU. A similar scheme occurs in RP-3, but the address spaces dedicated to local and shared memory are under complete software control, providing an all-local, all-shared or mixed memory access scheme.

Multifunction processing with PASM 227

PASM can emulate a shared memory system using a scheme very similar to that employed in the BBN Butterfly system. In the PASM prototype, however, there is no coprocessor to interpret whether the generated address is local or shared: the MC68010 CPU performs the actions of the coprocessor. Not having a coprocessor "between" the CPU and the local memory makes local references faster in PASM than in the Butterfly machine; however, the lack of a coprocessor to handle requests from other PEs forces the CPU to be interrupted to perform this function.

One way in which PASM can emulate a shared memory scheme is as follows. In each PASM PE's address space a certain range of addresses will be considered shared; generating an address in this range will cause local address-decoding hardware to generate a "bus error exception". During MC68010 exception processing, the internal state of the CPU is saved on the stack (even if mid-instruction) and the CPU begins to execute an exception-handling routine. In this case, the handling routine would be that for bus errors. Upon examination of the stack, the address that caused the bus error can be determined. If it is found to be an address in the shared memory area it is interpreted as a reference to a remote PE's memory and a message is generated and sent to the appropriate remote PE. When the message arrives at the remote PE it is interrupted, interprets the message and performs the transaction. If the transaction was a read the data is returned to the PE that requested the data. The original PE obtains the data, "patches" it into the stack (thus correcting the bus error due to the "non-existent" local data) and returns from the exception handler. The MC68010, which allows continuation from mid-instruction after an exception, now can proceed just as if the data item had come from its own memory. Of course, there is a performance penalty for the handling of each shared memory reference in software. Nonetheless, this scheme offers nearly as much flexibility in choosing the sizes of the shared data areas as does RP-3. Therefore no fixed size of the shared memory need be assumed by the hardware designer.

6 SUMMARY

The multifunction nature of the PASM parallel-processing system has been described. Its capabilities are derived from its ability to operate in either SIMD or MIMD mode and to switch between modes during execution of a program, to emulate a variety of machine topologies using a reconfigurable multistage network, to support multiple independent partitions of various sizes, and to allow both local and shared memory access. These capabilities make it a valuable tool for parallel architecture and algorithm research. In particular, the widely varying processing structures, types of data abstractions, and memory access patterns seen in the algorithms of image understanding tasks indicate that multifunction processing will be an important attribute of the next generation of parallel-processing architectures.

ACKNOWLEDGMENTS

This research was supported by the Rome Air Development Center under contract F30602-83-K-0119 and by an IBM Graduate Fellowship.

REFERENCES

[1] Batcher, K. E. (1982). Bit serial parallel processing systems. *IEEE Trans. Comp.* **31,** 377-384.
[2] Fountain, T. J. (1981). CLIP4: a progress report. In *Languages and Architectures for Image Processing* (ed. M. J. B. Duff and S. Levialdi), pp. 283-291. Academic Press, London.
[3] Uhr, L. (1981). Converging pyramids of arrays. In *Proc. IEEE Computer Society Workshop on Computer Architecture for Pattern Analysis and Image Database Management, November 1981*, pp. 31-34.
[4] Uhr, L. (1983). Pyramid multi-computer structures, and augmented pyramids. In *Computing Structures for Image Processing* (ed. M. J. B. Duff), pp. 95-112. Academic Press, London.
[5] Stolfo, S. J. and Miranker, D. P. (1984). DADO: a parallel processor for expert systems. In *Proc. 1984 Int. Conf. on Parallel Processing*, pp. 83-91.
[6] Gottlieb, A. Grishman, R., Kruskal, C. P., McAuliffe, K. P., Rudolph, L. and Snir, M. (1983). The NYU Ultracomputer—designing an MIMD shared-memory parallel computer. *IEEE Trans. Comp.* **32,** 175-189.
[7] Flynn, M. J. (1966). Very high-speed computing systems. *Proc. IEEE* **54,** 1901-1909.
[8] Bouknight, W. J., Denenberg, S. A., McIntryre, D. E., Randall, J. M., Sameh, A. H. and Slotnick, D. L. (1972). The Illiac IV system. *Proc. IEEE* **60,** 369-388.
[9] Batcher, K. E. (1977). STARAN series E. In *Proc. 1977 Int. Conf. on Parallel Processing*, pp. 140-143.
[10] Wulf, W. and Bell, C. (1972). C.mmp—a multi-miniprocessor. In *Proc. AFIPS 1972 Fall Joint Computer Conf.*, pp. 765-777.
[11] Swan, R. J., Fuller, S. and Siewiorek, D. P. (1977). Cm*: a modular multimicroprocessor. In *Proc. AFIPS 1977 Nat. Computer Conf.*, pp. 637-644.
[12] Cowther, W., Goodhue, J., Starr, E., Thomas, R., Williken, W. and Blackadar, T. (1985). Performance measurements on a 128-node Butterfly parallel processor. In *Proc. 1985 Int. Conf. on Parallel Processing*, pp. 531-540.
[13] Pfister, G. F., Brantley, W. C., George, D. A., Harvey, S. L., Kleinfelder, W. J., McAuliffe, K. P., Melton, E. A., Norton, V. A. and Weiss, J. (1985). The IBM Research Parallel Processor Prototype (RP3): introduction and architecture. In *Proc. 1985 Int. Conf. on Parallel Processing*, pp. 764-771.

[14] Kushner, T., Wu, A. Y. and Rosenfeld, A. (1981). Image processing on ZMOB. In *Proc. IEEE Computer Society Workshop on Computer Architecture Pattern Analysis and Image Database Management*, November 1981, pp. 88–95.
[15] Sejnowski, M. C., Upchurch, E. T., Kapur, R. N., Charlu, D. P. S. and Lipovski, G. J. (1980). An overview of the Texas Reconfigurable Array Computer. In *Proc. AFIPS 1980 Nat. Comp. Conf.*, pp. 631–641.
[16] Siegel, H. J., Siegel, L. J., Kemmerer, F. C., Mueller, P. T., Smalley, H. E. and Smith, S. D. (1981). PASM: a partitionable SIMD/MIMD system for image processing and pattern recognition. *IEEE Trans. Comp.* **30**, 934–947.
[17] Tanimoto, S. L. (1981). Towards hierarchical cellular logic: design considerations for Pyramid Machines. *Tech. Rep.* 81–02–01, Comp. Sci. Dept, Univ. Washington.
[18] Siegel, H. J., Schwederski, T., Davis, N. J. and Kuehn, J. T. (1984). PASM: a reconfigurable parallel system for image processing. In *Proc. Workshop on Algorithm-Guided Parallel Architectures for Automatic Target Recognition, July 1984*, pp. 263–291. [Also appears in the ACM SIGARCH newsletter: *Computer Architecture News*, **12**, No. 4 (September 1984), 7–19.]
[19] Siegel, H. J. (1985). *Interconnection Networks for Large-Scale Parallel Processing: Theory and Case Studies*. Lexington Books, Lexington, Massachusetts.
[20] Lawrie, D. H. (1975). Access and alignment of data in an array processor. *IEEE Trans. Comp.* **24**, 1145–1155.
[21] Meyer, D. G., Siegel, H. J., Schwederski, T., Davis, N. J. and Kuehn, J. T. (1985). The PASM parallel system prototype. In *Proc. IEEE Computer Society Spring Compcon 85*, pp. 429–434.
[22] Davis, N. J. and Siegel, H. J. (1985). The PASM prototype interconnection network. In *Proc. 1985 Nat. Computer Conf.*, pp. 183–190.
[23] Adams, G. B. and Siegel, H. J. (1982). The extra stage cube: a fault-tolerant interconnection network for supersystems. *IEEE Trans. Comp.* **31**, 443–454.
[24] Adams, G. B. and Siegel, H. J. (1984). Modifications to improve the fault tolerance of the extra stage cube interconnection network. In *Proc. 1984 Int. Conf. on Parallel Processing*, pp. 169–173.
[25] McMillen, R. J. (1982). *A study of multistage interconnection networks: design, distributed control, fault tolerance, and performance*. Ph.D. thesis, School of Electrical Engineering, Purdue University.
[26] Kuehn, J. T., Siegel, H. J. and Hallenbeck, P. D. (1982). Design and simulation of an MC68000-based multimicroprocessor system. In *Proc. 1982 Int. Conf. on Parallel Processing*, pp. 353–362.

Chapter Fourteen
Iconic and Symbolic Use of a Line Processor in Multilevel Structures

J.-L. Basille, P. Dalle and S. Castan

1 INTRODUCTION

Now that new parallel machines are reaching the fast expanding market of image processing, one should be asking what structures are best in relation to this field. In fact, all the general principles concerned with parallel structures have been understood for a long time and are well-known. As Fountain has said [1], perhaps it is time for university researchers to seek fresh channels of thought.

Therefore the task is to define, as accurately as we can, the scope of each parallel structure family in image processing. In this respect, we are putting forward two ideas in this chapter. The first is that a single structure cannot manage to solve all the problems found in image processing with sufficient efficiency and flexibility; it is therefore necessary to propose multistructure solutions.

Such a solution is all the more attractive since image processing comprises two domains: the iconic domain is concerned with all operations on the images in the true sense, that is, extracting features from the images; the symbolic domain is more concerned with pattern recognition and makes use of the description resulting from the previous stage (Fig. 1).

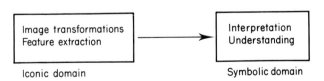

Fig. 1. *The two domains in image processing.*

The second idea is to show that, among the different structures that might be involved in multistructure systems, the line processor concept precisely provides a double solution, both to iconic and to symbolic processing.

2 WHY MULTILEVEL STRUCTURES?

Image processing is a very wide, spread-out field, which offers a rich variety of types and levels of processing. It is now very common practice to make a distinction between region-level and pixel-level processing [2]. Region-level processing necessarily leads us to MIMD structures. These structures may differ from each other with respect to their interconnection networks. In every case, two points need to be taken into account.

First, under certain assumptions, we note that limits exist to their range of use, depending upon the complexity of the processes involved and the

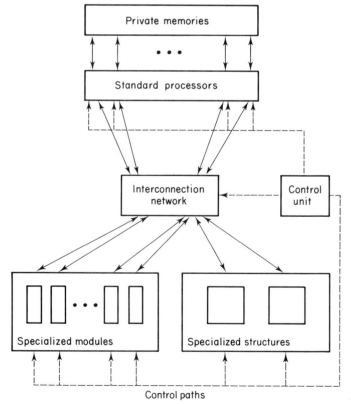

Fig. 2. A multistructure architecture for image processing.

throughput of the network [3]. The second point concerns the choice of an operating system, in order to avoid deadlocks. Strategies for resource allocation may be defined in a single-user context when, wherever possible, the programmer explicitly expresses the parallelism to be applied [4]. However, this level has been, up to now, less studied and is the more complex to deal with.

Pixel-level processing has been carried out in several structures. The two best known are pipelines and array processors. Systolic arrays and associative processors, for example, are also working at this level, but the advantage of pipelines and array processors lies in their flexibility and their appropriateness. This means that these structures are well matched to image processing, each for its own reasons, and can be programmed for the processing desired.

So a multistructure solution consists, on the one hand, of a MIMD part devoted to region-level processing and, on the other hand, of some specialized structure to perform pixel-level processing. But in order to complete the architecture and to improve its efficiency by adding cheap suitable tools, it is important to allow it to be connected to specialized modules such as, for example, a fast-Fourier-transform module. But other modules could be proposed for simple processes such as histogramming or, at the opposite extreme, a complex systolic array might be included.

Thus the complete architecture would appear as shown in Fig. 2. With regard to the specialized structure, we are inclined to favour array processors, but some questions arise which will now be discussed.

3 HOW MANY DIMENSIONS FOR AN ARRAY PROCESSOR?

When speaking of array processors one usually refers to bidimensional array processors. However, this term could be thought of as a broader generic term. It could include pyramids or, equally, line processors, as well as the more traditional bidimensional array processors. The question is then to decide whether it is better to select a one-, two- or three-dimensional array. Since images to be processed are commonly projections on a two-dimensional plane, bidimensional array processors are, of course, the most appropriate choice. Furthermore, when considering the reduction obtained throughout the successive processing stages, pyramids provide a structure that seems to fit the usual way of proceeding in image processing, from the lowest iconic level to the symbolic level.

But two aspects should be considered and cannot simply be ignored. The first point concerns efficiency. A good illustration can be obtained with processes involving only a part of the image. In this case, many of the processing elements of the array processor have to be masked and so relative efficiency decreases in proportion. Another problem arises in the case of pyramids, for which it is still particularly difficult to conceive a complete chain of algorithms intended to achieve pattern recognition.

The other point is cost. In spite of the constant progress in integrated circuits, the cost of an array processor is still very high and will always remain proportionately high. To give an idea, the cost of an ICL DAP, the first commercial array processor, which is only 64 × 64, was about £500000 in the early 1980s, in addition to the cost of the 2900 host computer [5]. Thus the use of array processors is limited to some huge and expensive systems, just when image processing is becoming more and more widespread.

In accordance with these two points, we have good reasons to believe that the line-processor concept is a satisfying middle road, and the new CLIP7 chip strengthens this standpoint. In particular, if we wish to design an image-processing machine at a reasonable price, compatible with a personal computer configuration, this is a good structure with respect to its low cost and its modularity.

A line processor is very modular, unlike some pipeline structures (which can be as cheap). If we do not possess enough processing elements to process an entire line of the image simultaneously, it is easy to carry out the processing of a line in several steps. It is therefore possible to define more or less powerful configurations which can be extended later on by simply connecting extra modules. For instance, we achieved the SIMD part (a line processor) of our project SYMPATI with only 16 processing elements, each having been designed with discrete TTL circuitry (Fig. 3). We are now planning to put each processing element on to silicon.

As with bidimensional array processors, a problem arises when the number of processing elements and the number of points in an image line are not equal. How can we manage the mapping between the image and the array? Two possibilities exist. We choose the most commonly adopted solution—interleaving. Some drawbacks to this solution have been pointed out, and the alternative, distributing the processors over the image, instead of distributing the image over the processors, has been supported [6]. But when all factors are taken into account, it appears that the interleaving solution is better. In particular it is better adapted to Input/Output, it makes it possible to take advantage of a global indicator, when, for instance, a segment of successive points belongs to a region of interest, and finally it makes it possible to implement propagation algorithms, as will be shown below.

A final question has to be examined. It may also be considered as a dimensional problem, although not in a spatial sense. It concerns the

Line processor in multilevel structures 235

Fig. 3. *The line processor, SIMD part, of the SYMPATI project.*

number of bits processed at each cycle. To date, nearly all array processors have been binary array processors, so that they work in a bit-serial fashion, which is of some interest, especially for numerical applications, when variable-length operands are considered. But this facility has to be costed against some extra work, and, particularly in image processing, we feel it is important to have a multibit processor structure. Once more the CLIP7 chip brings us encouraging corroboration for our initial choice. Nevertheless, if each pixel is coded on a byte, for example, intermediate results, inside the processing element, may need a double-length register and a double data path. We took this factor into account when designing the version of the processing element to be integrated.

4 AN ICONIC EXAMPLE OF THE USE OF A LINE PROCESSOR

In order to determine the distance between each pixel of an image component and the boundary of the component, many algorithms have been proposed. These algorithms are very important, as the results obtained, the boundary distances, are exploited in other algorithms to extract more synthetic features, such as skeletons, or to enhance the image, by erosion and dilation for example, or to perform other tasks such as data compression.

These algorithms are necessarily propagation algorithms. With a bi-dimensional array processor the propagation time will be proportional to the longest distance over the image. In any case, it will be rather fast, since all the distances are calculated simultaneously. At the opposite extreme, with a sequential monoprocessor, a classical algorithm provides the same result in a double scanning of the image. We shall see that this algorithm can be transformed so that all the pixels of a line may be treated in parallel. Then the algorithm obtained exactly matches a line-processor structure.

Let us first consider the algorithm used with a sequential mono-processor. $A(i,j)$ will refer to the pixel of line i and column j, whose value is initially 1 or 0 for pixels belonging or not belonging to an object, and at the end of the process, is the boundary distance. In the first top-down scan the lines are scanned from left to right. In the bottom-up scan they are scanned from right to left. The two respective treatments to be applied to each pixel are the following:

top-down scan
If $A(i,j) > 0$ Then
$A(i,j) \rightarrow \text{Min}\ (A(i-1,j-1), A(i-1,j), A(i-1,j+1), A(i,j-1)) + 1$
Endif

bottom-up scan
$A(i,j) \leftarrow \text{Min}\ (A(i,j)-1, A(i,j+1),\ A(i+1,j-1),\ A(i+1,j), A(i+1,j+1)) + 1$

The neighbourhoods respectively involved in each scan are shown in Fig. 4.

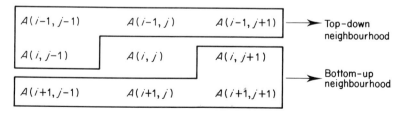

Fig. 4. *The neighbourhoods used in the sequential algorithm.*

Line processor in multilevel structures 237

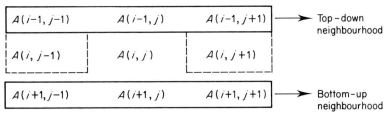

Fig. 5. *The neighbourhoods used in the line parallel algorithm.*

In order to transform this algorithm in the way proposed, we must take into account only the respective neighbourhoods shown in Fig. 5.

To perform the horizontal propagation we need only to test the presence of a 0-pixel on the left or on the right. These tests are done during the top-down scan. Thus the line algorithm also comprises two scans, processing each line in parallel, with the following processes:

top-down scan

If $A(i,j) > 0$ Then
$\quad A(i,j) \leftarrow \text{Min } (A(i-1,j-1), A(i-1,j), A(i-1,j+1)) + 1$
Endif
If $(A(i,j-1)) = 0$ Then $A(i,j) \leftarrow 1$ Endif
If $(A(i,j+1)) = 0$ Then $A(i,j) \leftarrow 1$ Endif

bottom-up scan

$A(i,j) \leftarrow \text{Min } (A(i,j)-1, A(i+1,j-1), A(i+1,j), A(i+1,j+1)) + 1$

It can be shown that the values supplied by this algorithm exactly correspond to the boundary distances of the pixels. The way to verify this is to consider the different conditions determining the boundary distances. Since the boundary distance of a pixel corresponds to the largest square, centred on the pixel and entirely contained in the object, the conditions determining the distances depend on the position of the 0-pixel closest to this square. At least one such 0-pixel exists. Figure 6 shows the different cases and some corresponding routes to the boundary.

5 SYMBOLIC USE OF A LINE PROCESSOR

Image processing ranges from simple iconic operations whose results, usually transformed images, are then used, directly or indirectly, by a human operator, to real image-understanding problems making use of both iconic and symbolic processes.

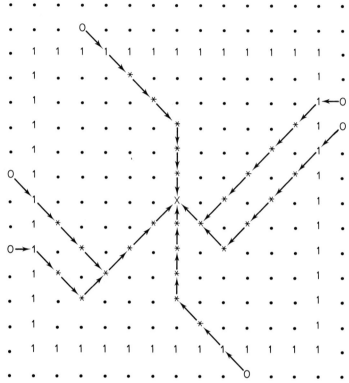

Fig. 6. The different ways to obtain boundary distances.

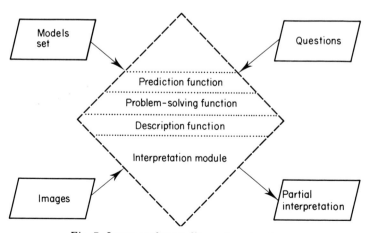

Fig. 7. Image-understanding system structure.

The image-understanding system we are developing [7] handles the following components (Fig. 7):

a set of scene models into which is put *a priori* knowledge of the images or the scenes to be analysed;
the images to be processed;
the questions to be answered;
the partial interpretation answering the questions;
the interpretation module, which must assume prediction, description and problem-solving functions.

In order to realize such an interpretation module, a rule-based system has been chosen [8]. The structure then consists of a production-rules base, a database which represents the interpretation status and a rule interpreter which assumes control and selects the rule to be executed (Fig. 8).

The rule interpreter works as follows: first it finds the competing rules which match some facts in the database; it then selects a rule from among this conflict set; finally it fires the rule selected, modifying the database.

As in other artificial-intelligence systems, an image-understanding system is usually written in LISP or PROLOG, which are time-consuming.

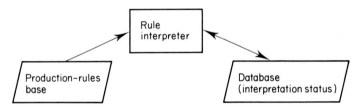

Fig. 8. Interpretation-module structure.

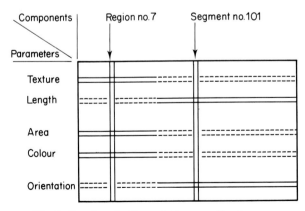

Fig. 9. Database implementation on a line processor.

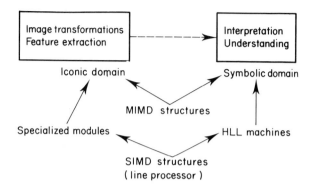

Fig. 10. *Structures for the two domains in image processing.*

So, to reduce the processing time, expensive high-level language (HLL) machines are often used, but the matching process is still very slow.

In order to propose a solution to this bottleneck, the matching could be carried out on a line processor. It would then only require the database to be structured as a column for each fact. A fact is the set of attributes attached to a component of an image. For example, a component may be a segment, a region, a surface or a volume. The attributes, that is, the values of the parameters, are ordered on the same line for a given parameter (Fig. 9). Alternatively, the facts may be of another kind: they may be rules. These are then filtered by meta-rules.

6 CONCLUSION

We have shown in this chapter that a line processor is relevant to both iconic and symbolic cases. It is therefore a good specialized structure to select for a multistructure solution, particularly for a low-cost configuration. Furthermore, multistructure architectures are also desirable in both iconic and symbolic cases. Thus Fig. 1 may be completed to give Fig. 10.

REFERENCES

[1] Fountain, T. J. (1983). A survey of bit-serial array processor circuits. In *Computing Structures for Image Processing* (ed. M. J. B. Duff) pp. 1–14. Academic Press, London.

[2] Basille, J. L., Castan, S. and Latil, J. Y. (1981). Système multiprocesseur adapté au traitement d'images. In *Languages and Architectures for Image Processing* (ed. M. J. B. Duff and S. Levialdi), pp. 205–213. Academic Press, London.

[3] Basille, J. L., Castan, S. and Al Rozz, M. (1983) Parallel architectures adapted to image processing, and their limits. In *Computing Structures for Image Processing* (ed. M. J. B. Duff) pp. 31–42. Academic Press, London.
[4] Basille, J.L., Castan, S. and Latil, J. Y. (1984). Structures parallèles en traitement d'images. *Premier Colloque Image CESTA, Biarritz, 21–25 mai 1984.*
[5] Hockney, R. W. and Jesshope, C. R. (1981). *Parallel Computers.* Adam Hilger, Bristol.
[6] Danielsson, P. E. (1984). Vices and virtues of image parallel machines. In *Digital Image Analysis* (ed. S. Levialdi). Pitman, New York.
[7] Dalle, P. and Castan, S. (1985). Système évolutif en analyse de scènes. *Rapport de Contrat* DRET 83/1052.
[8] Chambon, D., Debord, P., Dalle, P. and Castan, S. (1984). Configuration de base du système SACSO: interpretation de scènes à partir de modèles 2D. *4° Congrès RFIA AFCET, Paris, janvier 1984.*

Chapter Fifteen

Self-Learning Capabilities of VAP for Low-Level Vision

A. Favre

1 INTRODUCTION

Image processing is generally divided into low- and high-level categories. Low-level image processing usually consists of transformations modifying images into other images, whereas high-level transformations work on information already extracted from images. This classification hides the most important problem of information extraction (out of the image). These operations are sometimes gathered on an intermediate level. Before considering this intermediate level, it is worthwhile asking whether there is a limit to the application of low-level transformations and what are the characteristics of this limit, if any. In other words, we need a better understanding of what the local content of information could be.

We shall indicate the orientation we have chosen to investigate this kind of low-level problem and describe an elementary local image transform (low-level) which, in some sense, is optimal for dichotomic segmentation based on texture discrimination. This approach follows the classical supervised learning theory of Bayes [1,2] and yields a neighbourhood operation derived from co-occurrence matrices [3,4].

The prerequisite to these experiments is the VAP processor [5,6], which is used for on-line implementation of these image transformations. The comparison of this method with results obtained previously [6] "by hand" in the study of electron micrographs of liver cells will also be discussed.

2 THE VAP PROCESSOR

VAP has been designed and built to perform neighbourhood operations on-line with our Quantimet 720 system. The design originated from simulation studies of segmentation problems on electron micrographs, showing the possibility of locating different organelles of the cell by use of local image transforms [7].

VAP is divided into two parts (Fig. 1).

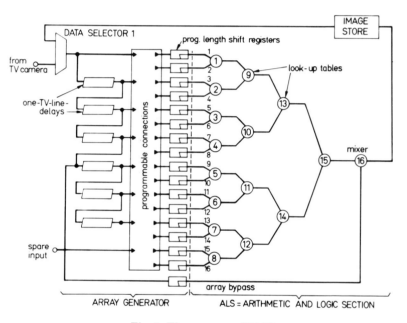

Fig. 1. The structure of VAP.

2.1 The arithmetic and logic section

On-line image transformations require pipelined computation, and the reduction of local information to one pixel value naturally extends the concept of a pipeline to the concept of a cascade. A binary cascade of PEs (processing elements) has been chosen for VAP. Each PE is a look-up table with two entries and one output. During the transformation of the image the two input values to the PE are combined to form a 12-bit address, and the 6-bit content of the table is output as the resulting value. Any function of two variables can thus be tabulated in each PE provided it can be expressed in 6-bit values.

Self-learning capabilities of VAP 245

2.2 The array generator

The sequence of pixels output by the scan (line by line) of the camera is partly retained in a system of delay lines and shift registers with programmable tap points giving a parallel access up to seventeen neighbouring pixels; sixteen of them are paired and used as input values to the eight PEs of the first level of the cascade, where eight partial transformations are computed in parallel. The eight results are then paired again and used as input values for the next level of four PEs, and so on. At the end of the cascade the result of the transformation can be compared with the original value (given by the 17th tap point) in an extra PE, and the final result is released from the processor.

3 THE METHOD

An important question is the choice of the functions to be tabulated in each PE of VAP. Our method, so far, has been to guess which transformation could be the best among the known collection of local operators and to adjust a few parameters such as various constants, threshold and the like so as to find a local optimum of the segmentation performance. We now try to get rid of the fixed algebraic formulation of these operators and compromise with the probabilistic characteristics of the neighbouring pixels' values.

3.1 The criterion

The following considerations apply to problems that could be solved in the general framework of Bayesian supervised learning. We assume that the objects to be localized in the images cover a "reasonable" part of the image area and also that a precise location should be achieved for further measurements. As a consequence we could estimate the validity of a segmentation algorithm by comparing the image of labels obtained (Object, Background) with the objects outlined with a light pen by a specialist. The misclassified area

Area(computed objects XOR outlined objects)

ranks the different possible algorithms.

This misclassified area is estimated by the number of pixels having the wrong label, and if the images are considered as stochastic processes this number is related to the probability of a wrong label. In order to minimize this probability, the Bayesian criterion is applied by computation of the likelihood ratio

P(label = "Background")/P(label = "Object")

for each pixel and the result thresholded to 1. If the ratio is greater than 1 then the best label of the pixel is "Background", otherwise it is "Object". A global optimization of the cascade seems difficult but the theory applies to each PE.

3.2 The content of the look-up tables (PE)

We shall first consider one PE of the first level of the cascade. Its two inputs consist of neighbouring pixel values of the original image. As the offset between these pixels is kept fixed during the scan, we shall use the local coordinate (0,0) for one pixel and (h,k) for the other. Their values are designated by [0,0] and [h,k] respectively and constitute the information available for the classification of the pixel (0,0). The thresholded likelihood ratio is needed to perform the classification, and in turn this requires the estimation of two probability distributions, namely

and P(label = "Object" AND pixel value(s))

P(label = "Background" AND pixel value(s))

Using the neighbourhood available, this can be rewritten as

and P(label = "Object" AND ([0,0],[h,k])

P(label = "Background" AND ([0,0],[h,k]))

The estimation of these distributions can be performed in various ways. For the sake of simplicity we shall assume an original image with only 3 bits per pixel. In this situation each occurrence of the pair [0,0] and [h,k] can be coded by a number between 0 and 63:

		[0,0]							
		0	1	2	3	4	5	6	7
	0	0	1	2	3	4	5	6	7
	1	8	9	10	11	12	13	14	15
[h,k]	2	16	.	.	.	.	.	.	.
	.	.	.	.	.	.	.	.	.
	.	.	.	.	.	.	.	.	.
	7	56	57	58	59	60	61	62	63

Self-learning capabilities of VAP 247

This table is loaded in the PE with input [0,0] and [h,k]. As a result, the transformed image displays the codes of the different co-occurrences found on the original image. Computing the area relative to each code allows the estimation of the probability of each co-occurrence of values (the co-occurrence matrix). This computation is split into two parts: one restricted to the outlined objects, the other to the background. Assuming that the objects outlined by a specialist have been stored as a binary image and that the area of a given code also produces a binary image, ANDing these two images will produce the desired results for the objects. The image of outlined objects is then negated and the corresponding result obtained for the background.

For the more general case of an original image with 6 bits per pixel, the content of the PE is reduced to the values $[0...6] \times [0...6]$ which are coded from 0 to 48, and the remaining part of the PE is filled with another code (e.g. 49). This is used for several steps, together with a look-up table (LUT) situated on the image path in front of the entry to VAP. At each step the content of the LUT is changed to map the successive intervals of seven values of the original image onto the values 0 to 6. All other intervals are mapped onto an extra value (e.g. 7). Once the areas of the different codes have been collected separately on the outlined objects and their background, the thresholded likelihood ratio is computed for each co-occurrence and the PE loaded with the corresponding values.

This result is optimal for a neighbourhood of two pixels, assuming that (h,k) is the best possible offset.

Two more questions must be answered to solve the problem for larger neighbourhoods.

One PE has been defined on the first level; can the other PEs of this level be filled with the same content?

Generally the answer will depend on the type of images to be processed; however, some comments are straightforward.

It is obvious that the method does not depend on a translation of the pair of pixels used as input for the PE. Consequently, if the second pair of input pixels differs from the first one by a translation the same content is used.

If the local information (given by two pixels) can be shown to be isotropic, the same content of PE can be used for pairs of pixels differing by translation and rotation. This could be the case with small line pieces spread with all possible orientations over the domain considered, regardless of the dependence existing between the two pairs of pixels in each neighbourhood.

The second question is how to proceed with the next level.

The same method can be applied with the following modification. Instead of thresholding the likelihood ratio, which gives a 1-bit result at the different PEs of the first level, a 6-bit information is kept by recoding the values of the ratio. We have chosen to maximize the entropy, and thus the "information", by dividing the range of values of the ratio into 64 equiprobable intervals each coded with its rank number. Once the PEs of the first level have been loaded with these partial transformations we can proceed to one PE of the second level by computing the probabilities relative to the co-occurrence of the two outputs given by the first level.

4 THE RESULTS

The segmentation of mitochondria in the liver cell by point grey-level discrimination leads to such high errors that any measurement would be meaningless. It appears, however, that neighbourhood operations significantly reduce these errors.

We have made an initial study, which has shown that a measure of the local edge content improves the segmentation of these organelles. The basic transformation

$$ABS([0,0] - [h,k])$$

has been used and applied with several offsets (h,k) in the first level of PEs. These various differences have been summed, giving the first transformed

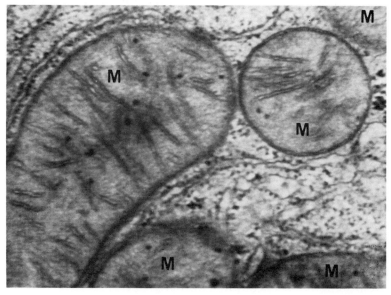

Fig. 2. Electron micrograph of a liver cell with mitochondria (M).

Fig. 3. The original image of Fig. 2 transformed by a local measure of the edge content.

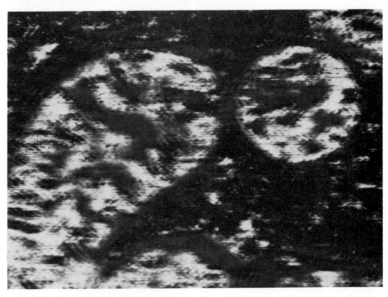

Fig. 4. The original image of Fig. 2 transformed by the transformation obtained by supervised learning.

image, which has then been smoothed. Figure 2 presents the original image. The image resulting after the transformation is shown in Fig. 3.

This study allows us to evaluate the results given by the new method where the different local operations are defined on the basis of the likelihood ratio as explained above. Figure 4 shows the images given by the new method; they correspond to those of Fig. 3.

The contents of the different levels of PEs are given in Fig. 5. Their precise algebraic definition is obviously impossible in terms of the usual arithmetic operations, but they correspond approximately to an inverted absolute value of difference for the first level and to a mean for the next three levels.

In essence the guesses we made during our first study were fair.

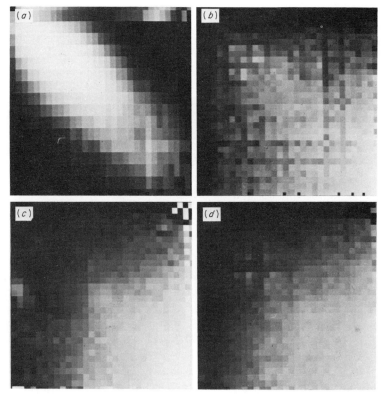

Fig. 5. The contents of the different PEs obtained by supervised learning: (a) level 1; (b) level 2; (c) level 3; (d) level 4.

5 DISCUSSION AND OPEN QUESTIONS

The segmentation technique presented here has two advantages: it is based on co-occurrence matrices, which guarantee certain properties of optimality, and it also produces a neighbourhood operation, which may be applied on-line.

The global algorithm, however, is suboptimal: the number of adjustable tabulated values is $N^2 (m-1)$, where m is the number of variables and N the number of values in the dynamic range. This can be compared with N^m for the general method. It is, however, difficult to appreciate the discrepancies between the two methods as the general algorithm is hard to implement for interesting applications. The method is related to the hardware structure of VAP and its binary cascade of two entries' look-up tables. Each table can be loaded at will, but it would have been too expensive to build a processor allowing the implementation of the general algorithm which requires an N^m memory.

Let us also recall the assumption of isotropy underlying the use of the same transformation in all tables of the same level: it means that the partial transformation loaded in the first level does not depend on the position or the orientations of the pairs of pixels taken as input for each table, but on their relative distance only. This argument applies to subsequent levels and the different groups of pixels involved at each entry of a table.

Rescaling the likelihood ratio onto the dynamic range needs some care in order to minimize the loss of information. It is questionable whether a transformation with equiprobable values brings a significant improvement as opposed to a logarithmic or linear rescaling of the ratio.

So far we must consider such an algorithm as a limit to local information extraction, when the assumptions for its application are fulfilled. In particular, unequal section thickness yields an illumination trend over all the image. The global modification of the local aspects of the texture due to this trend may have a bad influence on the algorithm performance.

Some problems are still unsolved.

The choice of the neighbourhood is a relatively complex question. It is not possible to determine the minimum neighbourhood, which gives sufficient local information to solve the problem, until "sufficient local information" has been defined.

In our first study we found a good length for the offsets between two pixels used as input for the PEs at the first level of VAP (with a given resolution) using the two covariograms estimated on the objects with respect to the background. If the two pixels are next to each other they bear values near to each other, and such a pair does not contain much more information than each of the pixels involved. On the other hand, long length offsets produce pairs with independent values, and it is questionable whether this improves the result of the segmentation. The same question arises at the subsequent level of PEs in VAP when grouping the partial results obtained as input of a table.

Another problem is to find a criterion to solve the special case of small objects (like lines) which cover 5% or less of the image area. One often considers it acceptable to make a 5% error during the segmentation of the objects. This amounts to saying that the original image covered with the label "Background" would be a good result, although probably not acceptable for line detection.

ACKNOWLEDGMENT

This work has been supported by the Swiss National Science Foundation, grant 3.524.83

REFERENCES

[1] Duda, R. O. and Hart, P. E. (1973). *Pattern Classification and Scene Analysis*, pp. 44–59. John Wiley, New York.
[2] McCormick, B. H. and Jayaramamurthy, S. N. (1975). A decision theory method for the analysis of texture. *Int. J. Comp. Inf. Sci.* **4**, 1–36.
[3] Haralick, R. M. (1976). Automatic remote sensor image processing. In *Digital Image Processing* (ed. A. Rosenfeld), pp. 47–52. Springer, Berlin.
[4] Zucker, S. W. and Terzopoulos, D. (1981). Finding structure in co-occurrence matrices for texture analysis. In *Image Modeling* (ed. A. Rosenfeld), pp. 423–445. Academic Press, London.
[5] Keller, HJ., Favre, A. and Comazzi, A. (1983). VAP—an array processor architecture using cascaded look-up tables. In *Computing Structures for Image Processing* (ed. M. J. B. Duff), pp. 143–157. Academic Press, London.
[6] Keller, HJ., Favre, A. and Comazzi, A. (1983). VAP—a video array processor using cascaded look-up tables and its applications in biomedicine. *Proc. SPIE* **397**, 406–414.
[7] Favre, A. and Keller, HJ. (1981). Local image transforms: an application in biomedical electron microscopy. *Patt. Recog.* **13**, 177–187.

Chapter Sixteen

The Impact of the Emerging Gallium Arsenide Integrated-Circuit Technology on Algorithms and Computer Architectures for Signal and Image Processing

B. K. Gilbert

1 INTRODUCTION

During the past fifteen years computer architects and systems designers have observed the continual and spectacular improvements in the device density and reliability of silicon integrated circuits with considerable excitement. The ability to place ever higher numbers of logic gates on a single integrated circuit has resulted in the proposal of complex computer architectures that would have been unthinkable even five years ago. Many such architectures have been described during several recent conferences devoted to this topic; all of these architectures exploit computational and functional parallelism (made possible by the nearly unlimited supply of logic gates typical of VLSI) to an unprecedented degree. Although some of the proposed architectures will probably not achieve their theoretical potential, a nontrivial number will almost certainly succeed. In so doing, these successful architectures will confirm the long standing promise of both VLSI and parallelism as a combined mechanism for solving complex computational problems.

During the past five years, however, an alternate integrated-circuit technology, digital gallium arsenide (GaAs), has been emerging from the research-laboratory phase into the glare of the popular (technical and nontechnical) press. Because of the higher effective mobility of the majority carriers (i.e. the electrons) in a GaAs crystal lattice than in a silicon lattice, logic gates fabricated on GaAs substrates exhibit faster switching speeds than do silicon logic gates, if all other device parameters remain roughly equal. In the first generation of GaAs gates, speed increases over silicon of roughly a factor of three are being observed; in second-generation GaAs structures, factors of ten over equivalent silicon transistors appear reasonable.

Notwithstanding these advantages of GaAs over silicon, at present the GaAs technology exhibits one serious drawback. Because GaAs fabrication process methods are significantly different from those of silicon, and perhaps fifteen years less mature, the number of GaAs gates that can be fabricated on a single chip is less than in silicon by an order of magnitude. Although no physical limits to the maximum number of GaAs transistors that can eventually be fabricated on a single chip have yet been identified, this disparate device-density relationship between silicon and GaAs is likely to persist for the foreseeable future; any percentage gains achieved in the gate counts of GaAs integrated circuits will be matched by equivalent improvements in silicon.

To exploit digital GaAs, then, the systems designer and architect must ask themselves the following question: if we have at our disposal a small number of extremely fast GaAs logic gates and small but fast GaAs memory components, how would we craft a processor architecture to exploit the strengths and avoid the weaknesses of the technology? This question, which has been explored at length over more than a decade for silicon, is only now being investigated for GaAs. This paper will describe several of the pertinent issues; however, the entire field is unexplored and is fertile territory for broad ranging theoretical studies.

2 GENERAL ARCHITECTURAL ISSUES

Specialists in the field of gallium arsenide integrated circuit technology have identified at least two generations of such devices; the first of these generations is based upon FET transistor structures roughly analogous to silicon MOSFETs; however, the GaAs FETs have demonstrated speed–power products from three to ten times better than silicon FETs of equivalent dimensions. These device characteristics translate into gate propagation delays in the range of 80–250 ps for loaded gates, and power dissipation levels of 0.1–0.2 mW per gate. Although legitimate differences of opinion have been expressed by the device-physics community, it appears clear that the devices are three to five times faster than their silicon equivalents. Further, a second generation of GaAs transistors has been identified, which is based upon the unique properties of sandwiches of gallium arsenide and gallium aluminium arsenide in alternating layers; these so-called heterojunction transistors can be created in both FET and true bipolar structures. The speed–power products of these second-generation transistors promise to be an order of magnitude better than that of the first-generation devices; FET structures have been demonstrated with gate delays in the 10 ps range [1], while bipolar ECL/CML devices have been demonstrated in the 20 ps range [2].

However, as noted earlier, although it appears that GaAs logic gates may be able to win the speed race with silicon, the latter technology is and will remain for the foreseeable future a much denser, higher-yield technology. Systems designers will have to use the speed of the GaAs gates to advantage, while avoiding the problems associated with low on-chip gate counts. The first lesson for the systems designer is that, because of the high on-chip gate speeds of this technology, a substantial penalty results whenever signals must be propagated from one integrated circuit to another. If the risetime for signals propagating on-chip is in the 20–100 ps range, but is degraded to 200–300 ps when it leaves the chip, and if the distances between two components on a circuit board is some multiple of one inch, then it can be stated approximately that every inch of separation between components is equivalent to compromising one or two ranks of logic gates in the system. It rapidly becomes apparent that, for large numbers of off-chip signal paths, the delays between components can equal the delays in the gates themselves; much of the potential performance in these fast gates can be dissipated in the interconnects if the systems designers do not account for these realities.

The solution to this problem lies in the selection of optimum algorithms for the implementation of any given logical or arithmetic function. For the case of silicon VLSI technology the most widely exploited option is the selection of algorithms with a large amount of on-chip parallelism because, as noted earlier, transistors are an inexpensive commodity in silicon. In GaAs, however, algorithms should be selected against a different set of criteria, three of which will be noted here. First, the algorithm should be implemented such that each data bit remains on a given integrated circuit for the maximum possible duration; if it is possible to "recirculate" the data bits from the outputs to the inputs of the chip for several cycles, as is characteristic of numerous iterative algorithmic functions, so much the better. Hence the available gates should be configured into deep and preferably pipelined types of implementations, not into broadly parallel structures. The use of pipelined architectures is particularly attractive because instruction and data pipes allow the use of higher system clock rates, which the GaAs gates tolerate extremely well.

The second criterion is a corollary of the first. Because even for signals propagating from gate-to-gate on a single chip a penalty is paid for the interconnect propagation delay time relative to the gate speed (a much smaller problem in many silicon technologies), the interconnect distances on each chip should be minimized wherever possible. The third criterion is a corollary of the first two, described above. Whenever it is necessary to communicate electrically between integrated circuits, the interconnects themselves must be treated as wideband transmission lines. Since in many cases the lines must be terminated in their characteristic impedance, and since these termination networks always dissipate power whether or not data is being transmitted, it follows that the system architects should attempt to arrange the processor data flow to minimize the number of lines (e.g. the widths of data busses) between components, between circuit boards, between subsystems, and between the nodes of a multiprocessor; further, the architects should attempt to maximize the data transmission rate on each data bit line. This procedure optimizes the utilization of the power committed to the termination networks.

3 LOGIC AND ARITHMETIC IMPLEMENTATIONS

With the design ground rules outlined in the previous paragraph, it is possible to identify classes of signal- and data-processing algorithms that will yield the best types of physical implementations if GaAs components are to be employed in the system assembly. Several of these mechanisms will be described in the following paragraphs.

It was noted earlier that the ability to employ algorithms that retain the data bits on each chip for the maximum possible duration, or that employ a small number of gates repetitively to subdivide a large problem into smaller problems through numerous recursive stages, will yield considerable performance benefits for GaAs-based implementations. One example of such an approach was presented by Swartzlander et al., as depicted in Fig. 1 and described in [3]. If it is desired to compute the inner product of two fixed-precision vectors, the classical approach has been to multiply the elements of the vectors pairwise, then sum the products. By expanding the equations for the inner-product operation to the bit-operand level, and then by rearranging the order in which the bit products and summations were taken, it was demonstrated to be feasible to create the inner product by a quasiserial approach with a much smaller number of gates than with a traditional parallel implementation. However, when performance measures were derived for the traditional methods and for that proposed by Swartzlander et al., it was found that the efficiencies of the implementations were very sensitive to the gate propagation delays specific to the transistor

QUASI SERIAL ARITHMETIC IMPLEMENTATION OF INNER PRODUCT COMPUTER

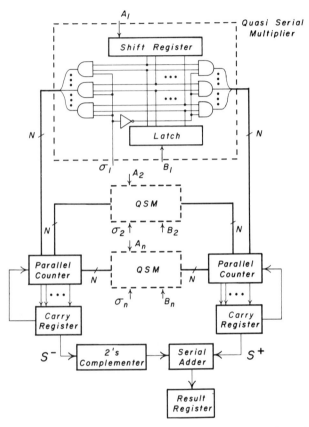

Fig. 1. An alternative, i.e. nonclassical, implementation of a fixed-precision inner-product operator. This implementation functions by processing the least significant bit of every element in both vectors in parallel, then processing the next most significant bit position in both vectors along with the carry terms from the rightmost column, and so on. The implementation uses a very low gate count, but must exploit very fast gates if it is to compete with classical implementations. (From [3] (© 1978 IEEE).)

technologies evaluated. In brief, unless the technology used to implement the quasiserial inner-product computer permitted gate propagation delays in the 100 ps range, the newer approach could not compete with the classical methods. Ten years ago, the classical methods were more efficient, since 100 ps gate speeds were not available. However, with modern GaAs technology, such approaches appear much more attractive and should be revisited. The original structures proposed by Swartzlander are in fact nearly optimum with regard to the criteria described earlier; that is, a small number of very fast gates are used over and over again to solve a large problem in a recursive rather than in a parallel manner.

As a second example of logic and arithmetic structures which should be able to exploit the best features of GaAs, the attention of the reader is called to the considerable amount of research that has been conducted (incidently, with silicon, rather than GaAs, in mind) in two generic areas: (i) connected nearest-neighbour array structures, such as the systolic array concepts presented by Kung, Leiserson and coworkers (see e.g. [4]), and recently extended by McWhirter, McCanny and coworkers [5]; and (ii) element-per-pixel array processors of the type exemplified by the CLIP series of image processing computers of Duff, Fountain and coworkers (see e.g. [6]). Kung and Leiserson have proposed that computers be assembled from identical, rather small operand-wide or word-wide computing elements which communicate only through nearest-neighbour connections. Unprocessed data operands would "flow" into the structure from its "top" and "sides", and be processed through consecutive processing elements on their passage across the array. McWhirter, McCanny and coworkers extended this concept by redefining the individual processing elements to operate on bit-wide streams rather than word-wide streams; in such an implementation the individual processing elements need only be of very low complexity, perhaps only 20–100 gates. The connections are nearest-neighbour only, as depicted in Fig. 2. The advantages of such structures for implementation in GaAs, aside from their low gate count and the repetitive nature of their on-chip layouts, are that none of the high-speed signals generated by the fast GaAs gates need propagate for more than a few hundred micrometres across the chip surface, and that each gate drives a signal to at most one other node in the system (thereby minimizing electrical loading and wavefront reflection problems). Finally, the individual processing elements can easily be designed to include pipeline latches at their inputs or outputs, which in turn allows the system clock rate to be increased to the limit of the gate technology; this approach directly exploits one of the major strengths of the GaAs technology, i.e. its rapidly responding gates.

In the approach proposed by Duff *et al.* in their CLIP series of engines, the interconnections of processing nodes are also nearest-neighbour structures; however, each node is somewhat more complex than in the systolic array structures, communicates more heavily with its nearest neighbours, and can perform a larger number of independent functions. Nonetheless, to a first level of approximation, the advantages of both types of structures are very similar for implementations in gallium arsenide: low gate counts, simple interconnects, reliance on fast gates and short propagation delays,

GaAs IC technology

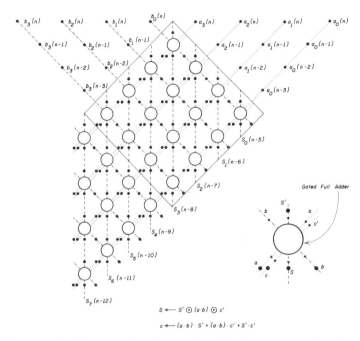

Fig. 2. An example of the types of implementation of fixed-precision arithmetic using bit-level systolic arrays. Each node in the processor is connected only to its nearest neighbours, and each is a very simple structure. In the example presented here the gated full adder node is comprised of only 25 gates, of which 18 are the adder and the remainder are pipelining flip–flops on the signal outputs. (From McCanny et al., Proc. 1982 Asilomar Conf. Par. Proc.)

and a minimum of interconnects all operating at the maximum bit rate per data line. As in the case of the bit-level systolic arrays of McWhirter and colleagues, the CLIP series of element-per-pixel processors can also be deeply pipelined to allow significant increases in system clock rates. Both of these types of architectures, and several closely related structures proposed during the past few years, are referred to as "cellular automata", because of the similarity of internal interconnection strategies and simplicity and identical layouts of the internal nodes of these processors. Many of these cellular structures, if proven to be feasible from an algorithmic execution standpoint, will be excellent candidates for implementation in high-speed logic technologies such as GaAs. For a more detailed review of these cellular automata structures the reader is referred to [7].

4 IMPLEMENTATION OF MEMORY COMPONENTS IN GaAs TECHNOLOGIES

To accrue the maximum advantage of GaAs transistors when considering memory component and memory page designs, it must again be noted that GaAs will offer very short access times but rather low total storage capacity per component when compared with silicon memory. For example, Kuroda and coworkers have presented data on a 4K-bit static RAM with a 2 ns access time, the fastest ever reported for RAMs of this size [8]. The fact that this RAM was among the first of the memory structures employing the new heterojunction FET structures of the type referred to as high-electron-mobility transistors (HEMTs), and that this second-generation technology is truly in its infancy, renders it strongly probable that much faster, though not necessarily much larger, GaAs RAMs will be forthcoming in the future.

For the systems designer, however, these new RAM devices pose a significant architectural problem. Although it is obvious that faster RAMs will always be useful for rapid-access scratchpads and instruction and data caches, merely fabricating such devices with conventional access and control strategies will not yield components that are useful for the systems designer. The reason for this paradox may be clarified by noting that if a traditional static RAM design with a cycle time of 20 ns is redesigned for an access time of, for example, 1 ns, the systems designer will be obliged to deliver address values and control pulses (chip select, chip enable, and write/read pulses) to these components with such precision, and of such short duration, that the mere propagation times on the memory circuit boards may make it impossible to satisfy the signal line constraints on the memory. For example, no responsible systems engineer would undertake the design of a memory board requiring the delivery of write pulses of 100 ps width; the ability of the system assembly technology to control pulse shapes and arrival times to such a high degree of accuracy is simply unequal to such an assignment.

There is, however, an approach to the design of short-cycle-time static RAMs that can in effect achieve the fast performance while leaving sufficient operating margin for the systems designer to integrate the memory with the computational portions of the system. As noted in the previous section, optimum designs for logic and arithmetic are likely to be deeply pipelined. Therefore, serious problems should not arise if the memory components themselves are designed with on-chip pipelining of address and data lines, and with many of the control signals (presently generated off-chip) included on the memory components. The RAMs will thus become synchronous devices, requiring the application of a clock signal to each chip. Two clock cycles might be required to execute a write operation and three cycles to execute a read function. For example, a 1 ns cycle time memory

GaAs IC technology

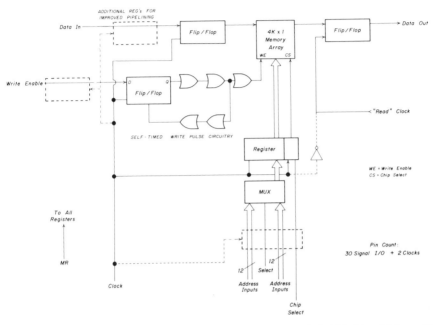

Fig. 3. One possible implementation of a gallium arsenide static RAM chip which would operate at approximately 1 ns cycle times, but which would allow the systems designers sufficient flexibility to exploit these high speeds. Note the inclusion on-chip of pipeline registers for input and output data bits and for the address, as well as two pairs of address lines. This latter function allows one set of addresses to undergo setup and propagation from their point of generation, while the other set of address lines is being applied to the memory array itself.

designed in this manner would provide the systems designer with 1 ns to set up the requisite memory addresses, 1 ns to fetch data from the internal memory array on the chips, and a final 1 ns to deliver the data to external logic components during three consecutive cycles of a 1 GHz clock. If a classical memory design were employed in a 1 GHz clock rate machine, the designer would have only 300 ps to execute each of the above described steps, a technically infeasible task. The internal structure of such a pipelined memory component is depicted in Fig. 3. For a further discussion of the problems that would have to be overcome if substeps of the memory cycle had to be executed in a few hundred ps see [9].

5 SUMMARY

The promise of gallium arsenide digital integrated circuits is widely believed to be in their potentially high operating speeds. This chapter has presented some of the problems that will have to be solved at the system-architectural level if these speeds are to be truly exploitable. We have reviewed the constraints on GaAs-based systems architectures, and the possible system structures that can circumvent the problems and deliver the promised performance of this emerging technology. Finally, it has been noted that all subsystem functions, i.e. the logic, the arithmetic, and even the memories of these processors, will have to be redesigned to deliver the full operational performance promised by the newest gallium arsenide digital structures.

ACKNOWLEDGMENTS

This research was sponsored in part by contracts MDA-903-84-C-0324 and F29601-84-C-0016 from the Defense Advanced Research Projects Agency. The author wishes to thank B. A. Naused, R. L. Thompson and D. J. Schwab, Mayo Foundation, for technical assistance; S. Roosild and S. Karp, Defense Advanced Research Projects Agency, for helpful discussions; and E. M. Doherty and S. J. Richardson for preparation of text and figures.

REFERENCES

[1] Pei, S., Shah, N., Hendel, R., Tu, C. and Dingle, R. (1984). Ultra high speed integrated circuits with selectively doped heterostructure transistors. In *Proc. 1984 IEEE GaAs IC Symp.; IEEE Publ.* 84CH2065-1, pp. 129–132.
[2] Asbeck, P., Miller, D., Anderson, R., Deming, R., Chen, R., Liechti, C. and Eisen, F. (1984). Application of heterojunction bipolar transistors to high speed, small-scale digital integrated circuits. In *Proc. 1984 IEEE GaAs IC Symp.; IEEE Publ.* 84CH2065-1, pp. 133–136.
[3] Swartzlander, E. E., Gilbert, B. K. and Reed, I. S. (1978). Inner product computers. *IEEE Trans. Comp.* **27**, 21–31.
[4] Leighton, T. and Leiserson, C. E. (1985). Wafer-scale integration of systolic arrays. *IEEE Trans. Comp.* **34**, 448–461.
[5] McWhirter, J. G., Wood, D., Wood, K., Evans, R. A., McCanny, J. V. and McCabe, P. H. (1985). Multibit convolution using a bit level systolic array. *IEEE Trans. Circuits Syst.* **32**, 95–99.
[6] Fountain, T. J. (1983). The development of the CLIP7 image processing system. *Patt. Recog. Lett.* **1**, 331–339.

[7] Preston, K. and Duff, M. J. B. (1984). *Modern Cellular Automata; Theory and Applications*, Chaps. 10 and 11. Plenum Press, New York.
[8] Kuroda, S., Mimura, T., Suzuki, M., Kobayashi, N., Nishiuchi, K., Shibatomi, A. and Abe, M. (1984). New device structure for 4K-bit HEMT SRAM. In *Proc. 1984 IEEE GaAs IC Symp.; IEEE Publ.* 84CH2065-1, pp. 125–128.
[9] Gilbert, B. K. (1985). Packaging and interconnection of GaAs digital integrated circuits. In *GaAs Microelectronics* (ed. N. G. Einspruch), Chap. 8. Academic Press, New York.

APPLICATIONS

Image processing has always been closely linked with medicine and biology, and some of the earliest specialized architectures were applied in these areas with great success. It therefore seems fitting that we should include the following two chapters, which describe projects in which multiprocessor systems have been used in medical image processing.

The first chapter describes the use of a laser scanner microscope and the Heidelberg Polyp MIMD multiprocessor in automated cytology, in a programme of research at the Optical Sciences Center of the University of Arizona (W. G. Griswold et al.). The Polyp comprises 30 PEs in the present system, and much hardware and software effort has been expended in order to make use of this very extensive computing power. In the second chapter Stephen Riederer, James Lee and Stuart Bobman give a helpful description of the way in which nuclear magnetic resonance is harnessed for medical imaging, and go on to describe their processor structure for rapid image construction and display—the only chapter in this volume concerned with image display rather than image analysis.

Chapter Seventeen

Multiprocessor Computer System for Medical Image Processing

W. G. Griswold, P. H. Bartels, R. L. Shoemaker, H. G. Bartels, R. Maenner and D. Hillman

1 INTRODUCTION

Computer-assisted diagnostic assessment of histopathological sections can provide the pathologist with valuable diagnostic and prognostic clues [1]. These might be based on quantitative measurements of micromorphometric features, on the quantitative evaluation of visually imperceptible criteria [2, 3] or on the results of cytochemical and histochemical photometric tests.

An effective system for computer-assisted diagnostic assessment requires both high-speed image acquisition and high-speed computer analysis of the image data. The ultrafast laser scanner microscope we are using to rapidly acquire high-resolution microscopic image data has been described in detail elsewhere [4–6].

The laser scanner operates in two modes. The continuous-scanning mode is used in applications where monolayers of separate individual cells are to be examined for the presence of abnormalities, for example for malignant or dysplastic cells, for cells showing signs of viral infection, toxic effects, or for cells labelled with specific markers. In this mode, a 2×2 cm^2 area is scanned at 0·5 μm resolution for a total of $1·6 \times 10^9$ pixels, with each point digitized to eight bits. The operating speed of the scanner is up to 6000 scan lines/second, with 4000 pixels/scan line in each of two wavelength channels. This gives a maximum digital data rate of 48 Mbytes/s/channel, with a 50% duty cycle. The total digital data rate is thus up to 96 Mbytes/s, and the total scan time is 80 s/slide. This data must be processed in real time, since it is not practical (or desirable) to store such a large amount of data. Fortunately, with isolated cells, scene segmentation and most other processing is fast enough to make real-time processing feasible.

The second scanning mode scans much smaller areas, a few square millimetres in area, with the same resolution and data rates being available. This scanning mode is used for the assessment of tissue sections. In this case, regions producing pixel volumes of the order of 10^8 pixels are of diagnostic interest, so the image data can be stored in the computer for subsequent analysis.

It is our purpose to present some of the considerations that went into the choice and design of the computer and operating system for doing image analysis of this sort. We will also introduce some other issues that arose in the course of our research. As a background for this discussion some of the basic problems faced in the assessment of histopathological images are presented first.

2 NATURE OF THE ANALYSIS

Histopathological sections offer imagery of considerable complexity, and while some pre-processing of the imagery at the pixel level is nearly always called for, the main challenge is posed by the need for consistent scene decomposition that leads to reliable extraction of the diagnostic information. For histopathological sections this diagnostic information is expressed primarily in the micromorphometric arrangement of the tissue components, i.e. of the larger scene components. To carry out such information extraction a knowledge database directed approach is required, with the knowledge database designed very specifically for the processing of sections from a given organ site.

It is important to realize that the recognition of diagnostic information in histopathological sections is very different from the problem faced by many machine-vision systems in robotics. There, strategies can be implemented that are directed at the recognition of a set of known man-made objects presented under different aspects, orientations, or under partial occlusion. The objects themselves, though, are rigidly definable in their morphology. This is not the case for the tissue components in histological sections.

The knowledge database for the processing of histopathological materials has to be organized in a two-layered structure. A topological description defines the invariant mutual relationships between different tissue components, and a topographical description provides a multivariate feature set, defining mean feature values, feature variance–covariance structure, and information about higher-order dependencies.

2.1 A parallel-processing approach

Both the nature of the algorithms for processing such images and the need for high-speed processing indicated to us that the use of a multiple-instruction-stream, multiple-data-stream (MIMD) multiprocessor computer system was the most suitable approach [7]. This led to an assessment of exactly how such processing could be done in a multiprocessor environment, and hence what the exact requirements for the system would be.

A practical example can best serve to illustrate our processing needs, and hence the needs for various capabilities of the computer and its operating system. Figure 1 shows a photomicrograph of a papillary adenoma of the

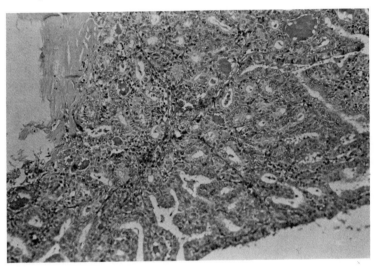

Fig. 1. Photomicrograph of a benign papillary adenoma of the thyroid.

thyroid [8], represented in Fig. 2 in greatly simplified form. The papillae are covered by a single layer of cuboidal cells, with ovoid shaped nuclei; stroma cells are few in number. There is, in this benign adenoma, almost no polymorphism of the cells. The structure shown in Fig. 1 has a diameter of approximately 500 μm; with a scan line spacing of 0·5 μm the extension of a 512 × 512 field as it would be acquired by a videophotometer is indicated in Fig. 2. This tissue section demonstrates some unique problems for processing. First, the diagnostic information is not uniformly distributed in the field. Secondly, a scene decomposition based on following a boundary might not be the best approach, since the boundary may not be contiguous:

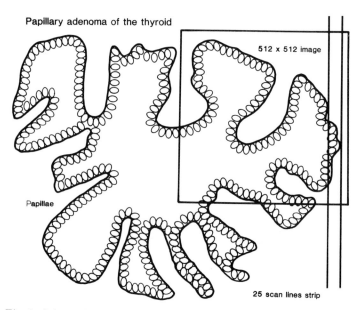

Fig. 2. Schematic drawing of the cuboidal cell layer in thyroid papillae.

rather a decomposition linking of all the dense nuclei of the cuboidal cell layer covering the papillae appears to be the best approach. This is also the decomposition done by a human observer. The diagnostic information is in the gross morphology of the papillae, as well as in the fine chromatin structure in the nuclei, as seen in Fig. 1. There is also diagnostic information in the total amount of nuclear staining. Feulgen-stained sections are used to establish the DNA ploidy pattern of the material, offering valuable prognostic information in case of adenocarcinoma of the thyroid [9–11]. Thus there is a need to find the single layer of cuboidal cells, trace the entire contour, delineate each nucleus, and process the high-resolution image of each nucleus for chromatin textural and cytochemical information.

Each of these tasks is best accomplished by rather different processing approaches, each suited to a different system configuration, as described below. For the following, we shall define a phase of an algorithm to be some large logical block of computation that may or may not be distributed across a number of processing elements (PEs). Phase X follows phase Y if Y's results are needed by X. This implies an ordering in time, also.

Phase I of the processing calls for a median smoothing algorithm over a 3×3 neighbourhood. Smoothing the image makes detection of the position of nuclei much more reliable, and needs to be performed before any further information extraction can proceed. This would best be done in parallel by a

Processing sequence thyroid tissue

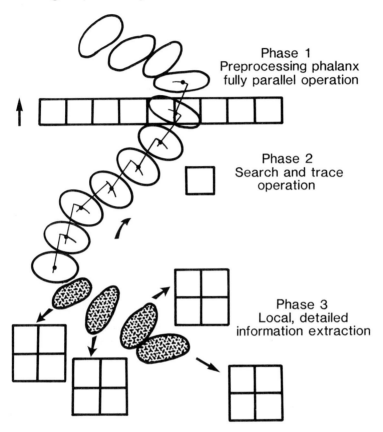

Fig. 3. *Schematic view of a processing sequence that could be used to analyse a thyroid tissue section. The squares represent groups of processing elements which work together.*

phalanx of PEs working independently as indicated in Fig. 3. This preprocessing is followed by grey-level histogramming, threshold determination, and use of the threshold value for creation of a binary image based on the original 8-bit image. The histogramming stage is performed by all available PEs, one region of the image per PE. This requires that averages from each region being scanned by a PE be collated by a single PE for final threshold determination.

The task of scene segmentation, phase II, illustrates the value of prior knowledge about the scene. The knowledge database for thyroid sections would provide the topological information that the papillae are bounded by a single layer of cuboidal cells, and that their centres are nearer to each other than centres of nuclei from stroma cells are. Connecting the centres of the cuboidal cell nuclei thus will lead to the desired segmentation. One does not have to rely on some boundary to be physically present and preserved without gaps and interruption.

For this processing task a single PE is suitable. The processing steps involve finding a cuboidal nucleus in the binary image, and determining its centre and the orientation of the long and short semiaxes. Then, a search sweep with an arm length a little longer than twice the average nuclear semiaxis is initiated, in the direction of the short semiaxis. If a neighbouring nucleus is detected the sequence is repeated: the nuclear centre is found, the long and short semiaxes are determined, and a new search sweep is begun.

All detected nuclei are listed with their centre coordinates and approximate areas, and tasks for individual cell feature extraction are generated to initiate phase III. The ability to queue these new tasks is critical, since phase II can be performed while the binary image is being generated by the last stage of phase I, so these tasks must wait until PEs are freed from phase I.

The tasks in phase III operate on the full 8-bit, two-wavelength images. They involve processing that can proceed in parallel for each nucleus, such as exact nuclear boundary tracing, extraction of the boundary chain code, and integration of the optical density values for each nucleus, computation of shape and orientation, extraction of values for nuclear chromatin texture and spatial distribution (such as peripheral tendency), and evaluation of spectral features of the nucleus.

These tasks could be carried out by several groups of say, four PEs each, working independently on nuclear images as they are released by the papillae tracing task. In this case, the four PEs are grouped in the sense that they are all working on the same cell. Other types of grouping are possible, such as using interprocess communications within a group to solve one indivisible task.

A single PE is assigned to the task of accumulating summary information about the entire scene when other higher priority tasks are not available for it to execute.

These processing needs prompted several considerations for the choice of a computer and design of the operating system.

3 THE POLYP SYSTEM

3.1 Hardware requirements

The processing flexibility demanded by the heterogeneity of these appli-

Multiprocessor system for medical image processing

cations imposes several requirements on the hardware of our multiprocessor system. The hardware of the system we chose, the Heidelberg Polyp [12–14] is well matched to our requirements. The Polyp is a fully distributed MIMD system, permitting fast parallel computation of independent tasks. It is constructed from a variable number of PEs (in our present system, 30) running Motorola 68000 CPUs, allowing for expansion. Any PE may contain a variable number of board level units, such as a CPU, a memory, a DMA controller, a cache memory, an error correcting controller, etc., enabling us to fit the needs of special image-processing problems that might arise later. The CPU boards can be easily redesigned to take advantage of newer processors. For example, processor boards that use the 68020 and 68881 floating point processor are currently being tested. The Polyp memory consists of a shared four gigabyte address space that is partitioned among the PEs, and connected by several (typically two to eight) 32-bit global data busses (see Fig. 4), making inter-PE data transfer fast and

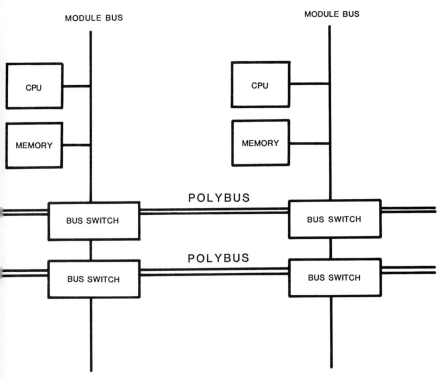

Fig. 4. Block diagram of the Polyp multiprocessor architecture. The number of the CPU/memory modules and the number of global data busses (the Polybus) can be varied as desired.

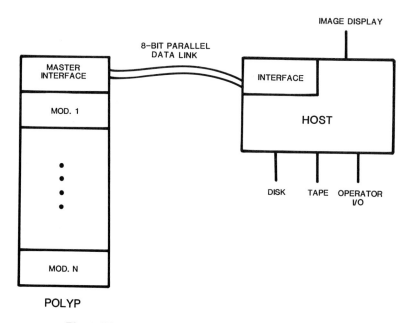

Fig. 5. Block diagram of the Polyp/host computer interface.

simple. The hardware allows run-time reconfiguration into PE "pools", i.e. groups of PEs addressable by one pool identifier. Finally, the addition of a Polyp Syncbus will permit very fast task assignment through hardware [14]. User I/O is performed via a host computer as shown in Fig. 5. To serve as a host to the Polyp system and aid in software development, we selected a Motorola 68000-based Codata 3400 supermicrocomputer employing a Multibus and running the UNIX* operating system. This type of system is ideal, since it employs the same processor as the Polyp, allows the use of a high-resolution imaging display and custom hardware boards, and has a very comprehensive set of languages and software development tools.

3.2 Software requirements

After the choice of the Polyp hardware and the host, attention focused on the operating system. Since disk accesses to the host computer are relatively slow, and the total amount of program code for our applications is not excessively large, it is possible to load all programs needed to analyse a particular image before run time. This both enhances the system performance and simplifies the demands on the operating system. Further, we

*UNIX is a trademark of A. T. & T. Bell Laboratories.

decided (for now) that since all tasks must eventually be completed, we would initially let each task run to completion without interruption. The problem of task management is discussed further in Section 3.3.1. Device waiting in the system is minimized by using an intelligent I/O server and suitable parallel processing primitives.

Because the system is data driven, and data sets can be widely varied, the operating system needs to permit internal dynamic task creation, and be elastic enough to handle swells of processes, queueing them until busy PEs become available. As demonstrated in the example in Section 2.1, the ability to dynamically create groups of PEs working together to solve a problem, using a set of interprocess communication primitives is essential. Following Comer [15], we define a process to be a computation that is performed on a single PE. In contrast, a task may be completed by several processes working together. Following the naming conventions for nodes in trees, a process created by another process is called a child.

3.3 The Polyp operating system

An operating system for the Polyp has been written at the University of Arizona, and provides the basic services of interprocess communication, process creation and invocation, Polyp-host I/O and memory management. This system is implemented in C, except for a few critical routines, which were written in 68000 assembly language. The underlying structure of all Polyp communications is message passing implemented with queues. Memory management is first-fit with global garbage collection. The assignment of new processes is managed through a scheme called contention allocation (see Section 3.3.1). To execute a program, one PE executes an initial program that spawns processes to be run on other PEs. These in turn can spawn additional processes, thus creating a process tree. Taking advantage of the Polyp architecture, the system is fully distributed, providing a level of fault tolerance. The system in its current format assumes homogeneity: all PEs have the same operating system and user program. In addition, there are a few global system tables that reside in one of the PEs. Hardware pooling as described in Section 3.1 can relieve this restriction. The system is fully operational in the form described in this paper.

All programs to be run on the Polyp are developed and compiled on the host, using the standard Unix C, FORTRAN, or PASCAL compilers and a modified library fitting the needs of the Polyp system. Polyp operating system code is kept separate from user Polyp programs, and the system code is available to the user via a trap interface. The Polyp operating system is passive in the sense that it only executes some initialization code and then stops. It performs no services not directly invoked by the user.

The other segment of the Polyp operating system runs on the Codata, which handles the loading and starting of programs, and monitoring for I/O requests. These operate through a set of protocols dictated by the hardware interface, and use a minimal set of hardware dependent primitives written in C to communicate with the Polyp.

3.3.1 Task management

Of particular relevance for the design of a multiprocessor operating system is the scheduling of subtasks of a program. A scheduling algorithm, even if efficient for most types of programs, can adversely affect the performance of other types. It became apparent that it might be beneficial for the programmer to be able to specify his own strategy, or to have the strategy based on a knowledge database of image features and processing times for those features.

Task scheduling can follow many different strategies, with varying efficiency and complexity. In a typical situation the majority of PEs would be assigned to one of several groups of PEs, each occupied with an assigned task. At the same time, a number of pending tasks might be queued. It is useful to distinguish between two kinds of tasks: tasks that must precede further information extraction, such as filtering operations or determinations of a set of locations which will serve as starting points for further processing steps; and tasks that cannot proceed until the abovementioned processing is completed. Let it be assumed for the following discussion that all pending tasks could proceed as soon as enough PEs become available.

In a first-in-first-out (FIFO) strategy, the oldest pending task, requiring say, two PEs, is activated as soon as at least two PEs are released, even if the just completed task releasing the PEs provides more than is needed. This will guarantee that all tasks will be served eventually, but could leave a number of PEs idle for a time, particularly if the pending task requires a large number of PEs, that at the time are not available. This strategy is easy to implement and reliable, but under certain circumstances wasteful of processing capacity.

A best-fit strategy considers all pending tasks in the queue, and always tries to engage the greatest number of PEs over the longest possible time span, rather than at a particular instant of task assignment. This means that smaller tasks might be temporarily postponed, and larger tasks preferred. More advanced versions of the best-fit strategy could employ a look-ahead schedule and thus provide optimum utilization of capacity. This is an impossible strategy to implement in general, since the performance depends

Multiprocessor system for medical image processing 277

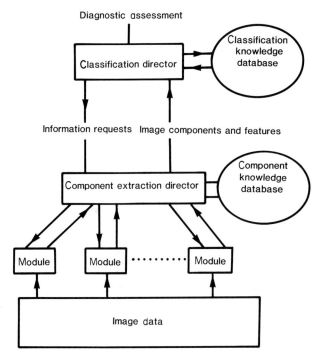

Fig. 6. *A possible hierarchical structure for a knowledge-base-controlled image analysis program.*

strongly on the task size distribution of the processing problem, as well as on the run times of the individual tasks. This, however, is information which is available in a knowledge-base-controlled processing sequence. One possible hierarchical structure for such a knowledge-base-controlled image-analysis program is shown in Fig. 6. The processing requirements for tissues from a given organ site are in fact very well known, and this extensive prior knowledge can be used to optimize throughput. It is known in what order different tasks will be invoked and what the task sizes are (both with respect to the number of PEs required for the task and to the expected run times). In this circumstance the database designer and programmer can exert major control by defining tasks and processes in such a manner that system capacity is always utilized to the greatest extent.

Task management strategies thus present a very different problem for a knowledge-database-controlled system than for the general case where optimization of throughput is the goal, without prior knowledge of task sequence, task size distribution, and expected run times. But with such prior knowledge, throughput optimization has to rely less on the complexity of the task management software and may take advantage of the ability to control task definitions.

The definition of such a knowledge-based heuristic is not trivial, so until a workable heuristic is developed, a FIFO strategy based on contention allocation has been adopted. Under this strategy, a process creating a new process puts a record of this process on a system queue, and proceeds on its way. Any PE that is idle removes a new process from a system queue and activates it. A priority strategy can be imposed on a process by placing it on a system process queue that is associated with the priority assigned to the process. Contention allocation offers several advantages. It incurs no system overhead, since a task watchdog is not necessary: task assignment is performed by a PE that is idle. Also, PE grouping is automatic using contention allocation. A process needing to group n PEs need only create n processes designed to work together (see Section 3.3.2). The first n PEs that become available pick up the created processes and invoke them. On the other hand, $n-1$ of the n processes might be invoked, and then wait for some indefinite time for the last process to be picked up. Avoiding such problems is the responsibility of the programmer. Ultimately, the success or failure of contention allocation compared to a best-fit strategy depends upon the time lost to non-optimal scheduling versus scheduling overhead.

This handling of new processes is done via a set of system primitives. A new process is created and put on a system process queue by the call

$$\text{spawn}(\text{function},\text{priority},\text{nargs},\text{arg}_1,\text{arg}_2,\ldots,\text{arg}_{\text{nargs}}),$$

in which **function** is the function (process) name, **priority** is the priority of the process in the task allocation strategy, **nargs** is the number of arguments to the new process, and **arg**$_i$ is the ith argument in the argument list of the function. If for some reason **spawn** should fail (i.e. all process queues full), a failure signal is returned. This is usually a panic condition. The **spawn** operation may be split into two logically distinct parts, the creation of the process, by

$$\text{process} = \text{createchild}(\text{function},\text{priority},\text{nargs},\text{arg}_1,\text{arg}_2,\ldots,$$
$$\text{arg}_{\text{nargs}})$$

and the sending of the process to a process queue, by

$$\text{status} = \text{sendchild}(\text{process})$$

which returns to status the success or failure condition of the action.

This division of the spawn operation permits the dynamic creation of a task group by contention allocation. First the group of processes is created by **createchild**, and stored in an array. Next the call

$$\text{status} = \text{sendfamily}(\text{nproc},\text{procarray})$$

is made, which atomically places the **nproc** processes on the system process queue, using the priority of the first process as the priority tag. This operation must be indivisible because two large groups might be created at once by two processes, and if they get mixed together, too many PEs might be demanded at once, causing deadlock.

The receipt and invocation of processes is divided between two primitives. The first,

process = getchild()

assigns to **process** a record of a process from a process queue. If there are no processes waiting to be invoked, **getchild** returns a null process. The second primitive is

invokechild(process)

which begins execution of **process**, a process received by **getchild**. The separation of the actions of **getchild** and **invoke** permits relaxation of synchronization [16] by allowing a PE failing to receive a process to perform lower priority "in PE" tasks such as writing out accumulating output, if no processes are available from the system.

As an example of assignment of PEs to processes for a 30 PE system see Figs 7 and 8, which are different views of the same tree generated by the example in Section 2.1. Note that the process tree is uniquely determined for a particular algorithm and data set, but the assignment of a PE to a process (a node in the tree) is based upon the availability of PEs, and a PE is reassigned to a new process (node) upon completing a process. Thus it is not possible (or necessary) to know in advance which PE is executing a particular process. Also note that some of the process transitions are internal, meaning a process not created, but simply called directly as a procedure.

Phase I is begun by a single PE being automatically assigned the main program, which reads in the image and partitions it into 30 regions. The main program creates 29 processes (leaving one region for itself) by repeated calls of

process[i] = createchild(phase1,0,3,region[i],inQ,outQ)

in which **i** takes the values from 0 to 28, and then sends them to the system atomically with the call

sendfamily(29,process)

The 29 idle processors wake up and perform

invokechild(getchild())

to start up the waiting processes. **InQ** and **outQ** are queues used for

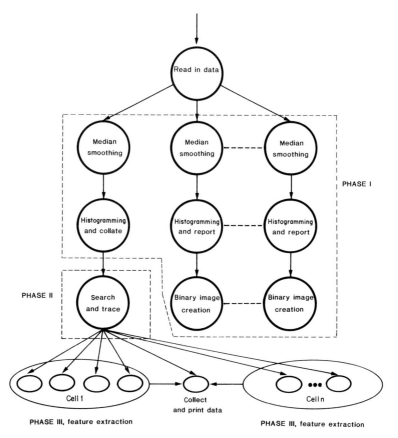

Fig. 7. View of a process tree that might be generated during the processing of the thyroid papillae image.

reporting of the threshold (see Section 3.3.2). Phase II is automatically entered by the PE executing the main program, and then initiates phase III in the course of locating cells.

3.3.2 Interprocess communication

Another important aspect of a multiprocessor system is its interprocess communications primitives, which are used by both the operating system and the user. Both their semantics and implementation influence what a programmer will want to do. Currently it is widely accepted that the best primitives for interprocess communication are based on message passing

Multiprocessor system for medical image processing 281

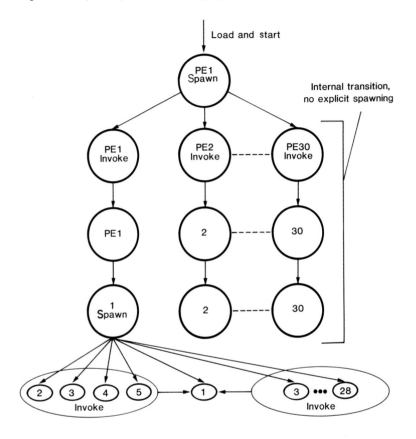

PE assignment tree

Fig. 8. A second view of the process tree for thyroid papillae analysis, showing a possible assignment of PEs to processes that might occur during a given run of the program.

[17,18]. While message passing is nearly imperative in a multiprocessor system without shared memory, it is still desirable in a shared memory system for several reasons. A message passing action is explicit and atomic (e.g. send or receive), thus fitting nicely in the semantics of a sequential language like C or PASCAL. Other interprocess communication primitives can be implemented with message passing, demonstrating its generality [18].

Message passing in the Polyp system is implemented using queues. This approach was chosen because queues allow a relaxation of synchronization by permitting incoming messages to queue up before old ones are received.

Also, unlike many other approaches to message passing [18,19], messages are not sent directly to processes or "objects", but to intermediate holding places (queues) that are objects themselves, and of which a process must be explicitly aware. Queues, like other data objects in C, can be passed as arguments to functions, and hence as arguments to new processes, making the creation of explicit interprocess communication links a natural part of the program. Thus the grouping of processes by sharing of queues, or the isolation of a process by not sharing of queues, is automatic, explicit and general. The basic transmit primitive in the Polyp operating system is

$$status = send(Q, message)$$

which sends **message** to **Q** if **Q** is not full, in which case it returns a failure condition to **status**. Conversely

$$status = receive(Q, message)$$

gets a message from **Q** and assigns it to **message** if Q is not empty, in which case **receive** returns a failure condition to **status**. The amount of synchronization required with a message queue is a function of the algorithm, the input data, and the length of the queue in use. So to specify the length of a queue

$$Q = newq(maxnum)$$

allocates a queue **Q** capable of holding **maxnum** messages.

As an example of interprocess communication, we can use the collation of the grey level threshold values in phase I of the example in Section 2.1. It is necessary for each process performing histogramming to report its threshold value to one process to have the average threshold of the image computed. This is accomplished by having each process send its value to the queue **inQ** by

$$send(inQ, threshold)$$

and then having the collating process remove all the values by 29 successful calls of

$$receive(inQ), threshold$$

After the thresholds are averaged, a single value is returned by

$$send(outQ, avg\_threshold)$$

as shown in Fig. 9(*a*), after which each histogramming process removes this value by calling

$$peek(outQ, avg\_threshold)$$

which looks at the oldest (and in this case only) message on **outQ**, but does not remove it, as shown in Fig. 9(*b*).

Multiprocessor system for medical image processing

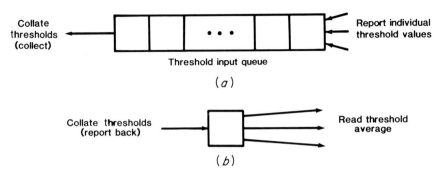

Fig. 9. The queue structures used for the determination of a threshold value for an image. (a) 29 processes send a threshold value for sections of the image to a 29-element queue. Another process receives all the values on the queue and (b) sends the average value back via a single-element queue which is examined by the other 29 processes.

3.3.3 Polyp input–output

Once a Polyp program has begun, the Codata becomes the I/O server for the Polyp. A process makes a request by putting a record of the request on the system I/O queue. The Codata then reads the request from the queue. I/O requests are handled strictly on a FIFO basis. The execution stage is carried out completely by the Codata, which reads and writes in the Polyp's memory whatever information is necessary, and calls the appropriate UNIX routines. This is transparent to the user, since all standard I/O routines follow the protocols of the UNIX operating system. However, in the attempt to increase parallelization, buffered writes are not waited on by the PE, but are the responsibility of the Codata. This enhancement is invisible to the user. Similarly, it is possible to request a read be done without waiting for it to complete until the input data is actually needed. The read enhancement is not invisible to the user, but it is an extension to the standard system, and does not have to be used.

3.3.4 Memory management

Because the Polyp hardware is based on a shared memory model, the handling of dynamic memory allocation is more complicated than in the uniprocessor case. However, the memory space is explicitly partitioned among the PEs. Thus it was natural to have each PE manage its own part of the memory space, using first-fit protocol in our case. The only problem with this approach was handling the freeing of memory blocks that are loaned as arguments for a spawned process or through message passing to another process. This is handled by a garbage collector in each PE that maintains in-use lists of memory. When a loaned piece of memory is freed, the block is marked free, and the PE that owns the block reclaims it by periodically searching the PE's in-use list looking for freed blocks, and returns any found to the free list. The user protocol for memory management is entirely compatible with UNIX.

4 FURTHER IMPLICATIONS FOR IMAGE ANALYSIS

If certain assumptions are satisfied, the algorithm described in Section 2.1 is inherently efficient. The granularity of the tasks is large, making the overhead for process start-up insignificant. Most processes are completely independent from others, and for the remaining processes a minimum of interprocess communication is required, implying that synchronization (waiting for a message response) is minimal. The only synchronization overhead in this operation is the pausing of 29 of the PEs while waiting for the threshold value, and the waiting of the phase II PE for the first complete cell to appear in the binary image. The waiting of a PE for processes is minimal, since the location of a cell in phase II is fast, while the extraction of features in phase III is much slower. The choice of four PEs to perform the cell feature extraction, instead of perhaps one, is made on the assumption that much less than 29 cells, but approximately 29/4 cells, can be located in the time it takes to create the binary image. Hence a greater number of PEs are mobilized immediately for phase III when phase I completes. There is perhaps some sacrifice in efficiency if coordination among the processes is necessary, but the result is overall speed-up due to high PE utilization in this early stage [20]. If there were more cells in a tissue section, a switch to one PE doing all feature extraction for a cell might be desirable once all PEs were mobilized [20]. Determining exactly how to get maximum utilization of the PEs during phase transitions is tedious, since it requires knowing approximate execution times for several computations, the amount of overhead that arises from a particular partition (i.e. interprocess synchronization [17] and phase start-up to full mobilization), how changing the partition of one phase affects other phases, and how much data a phase has to process. It is exactly this type of problem that could be best solved using a knowledge database, rather than by brute-force benchmarking on behalf of the programmer, since the amount of necessary information can be overwhelming, and its processing computationally intensive.

This use of the database is markedly different from its use in best-fit scheduling, because here the database is used in writing the program, rather than managing it while it runs. The formation of this database and its application would not be trivial. Also, indeterminate variables such as hardware interaction present a problem. However, the automation of optimizing perhaps the most critical factors in algorithm speed-up—proper partitioning for smooth phase transitions and minimal synchronization—still appears to be a worthy goal.

DISCUSSION

While many applications in high-speed image analysis can require specialized machinery to perform efficient pixel-by-pixel processing, histopathological analysis of cell samples and tissue sections requires a higher-level, more flexible approach. The use of an MIMD computer with a flexible, distributed operating system, capable of performing any type of analysis, and being dynamically reconfigurable seems paramount. While writing parallel algorithms for the Polyp system is not difficult, determining the most efficient partition of an algorithm among several PEs can be a tedious task. In the future, even faster and more efficient and accurate image analysis should be attainable by the use of knowledge databases to direct not only image processing and task scheduling, but also task partitioning.

ACKNOWLEDGMENT

The work described in this chapter has been supported by the National Institute of Health, grant 1-R01-CA34830-01.

REFERENCES

[1] Baak, F. P. and Oort, F. (1983). *Morphometry in Diagnostic Pathology*. Springer, Berlin.
[2] Bartels, P. H. and Wied, G. L. (1981). Automated image analysis in clinical pathology. *Am. J. Clin. Path.* **75,** 489.
[3] Bartels, P. H., Olson, G. B., Lockart, R. and Wied, G. L. (1980). Cytomorphometric studies of cell populations. *Cell Biophys.* **2,** 339–351.
[4] Bartels, P. H., Buchroeder, R. A., Hillman, D., Jonas, J. A., Kessler, D., Shoemaker, R. L., Shack, R. V., Towner, D. and Vukobratovich, D. (1981). Ultrafast laser scanner microscope, design and construction. *Anal. Quant. Cytol. J.* **3,** 55–66.
[5] Shoemaker, R. L., Bartels, P. H., Hillman, D., Jonas, J., Kessler, D., Shack, R. V. and Vukobratovich, D. (1982). An ultrafast laser scanner microscope for digital image analysis. *IEEE Trans. Biomed. Engrg.* **29,** 82–91.

[6] Shack, R. V., Bell, B., Kingston, R., Landesman, A., Shoemaker, R. L., Vukobratovich, D. and Bartels, P. H. (1982). Ultrafast laser scanner microscope—first performance tests. In *Proc. Int. Workshop on Physics and Engineering in Medical Imaging*, pp. 49–57. IEEE Computer Soc. Press, IEEE Catalog No. 82CH1751-7.
[7] Bartels, P. H., Maenner, R., Shoemaker, R. L., Paplanus, S. and Graham, A. (1983). Computer configurations for the processing of diagnostic imagery in histopathology. In *Computing Structures for Image Processing* (ed. M. J. B. Duff). Academic Press, London.
[8] Droese, M. (1980). *Cytological Aspiration Biopsy of the Thyroid Gland*. F. K. Schattauer, Stuttgart.
[9] Sprenger, E., Loewhagen, T. and Vogt-Schaden, M. (1977). Differential diagnosis between follicular adenoma and follicular carcinoma of the thyroid by nuclear DNA determination. *Acta Cytol.* **21**, 528–530.
[10] Lukacs, G. L., Balazs, G. Y. and Zs-Nagy, I. (1979). Cytofluorimetric measurements on the DNA contents of tumor cells in human thyroid gland. *J. Cancer Res. Clin. Oncol.* **95**, 265–271.
[11] Johannessen, J. V., Sobrinho-Simoes, M., Tangen, K. O. and Lindmo, T. (1981). A flow cytometric deoxyribonucleic acid analysis of papillary thyroid carcinoma. *Lab. Invest.* **45**, 336–341.
[12] Maenner, R. and Deluigi, B. (1981). The Heidelberg Polyp—a flexible and fault-tolerant polyprocessor. *Comp. Phys. Commun.* **22**, 279–284.
[13] Maenner, R., Deluigi, B., Saaler, W., Sauer, T. and Walter, P. V. (1984). The Polybus: a flexible and fault-tolerant multiprocessor interconnection. *Interfaces in Comp.* **2**, 45–68.
[14] Maenner, R. (1984). Hardware task-processor scheduling in a polyprocessor environment. *IEEE Trans. Comp.* **33**, 626–636.
[15] Comer, D. (1984). *Operating System Design: Xinu Approach*. Prentice-Hall, Englewood Cliffs, New Jersey.
[16] Jones, A. K. and Schwarz, P. (1980). Experience using multiprocessor systems—a status report. *Comp. Surveys* **12**, 121–165.
[17] Williams, E. (1983). Assigning processes to processors in distributed systems. In *Proc. 1983 Int. Conf. on Parallel Processing*. pp. 404–406, IEEE Comput. Soc. Press.
[18] Marovac, N. (1983). On interprocess interaction in distributed architectures. *Comp. Arch. News* **11**, no. 4, 17–22.
[19] O'Leary, D. P. and Stewart, G. W. (1985). Data-flow algorithms for parallel matrix computations. *Commun. ACM* **28**, 840–853.
[20] Chen, J., Dagless, E. L. and Guo, Y. (1984) Performance measurements of scheduling strategies and parallel algorithms for multiprocessor quick sort. *IEE E Proc.* **131**, 45–54.

Chapter Eighteen

Processing Techniques for Magnetic Resonance Image Synthesis

S. J. Riederer, J. N. Lee and S. A. Bobman

1 INTRODUCTION

Nuclear magnetic resonance is a phenomenon that was discovered in the late 1940s independently by Edwin Purcell and Felix Bloch [1,2]. Owing to the intrinsic spin angular momentum, certain nuclei such as protons, when placed in a magnetic field, develop a net magnetization aligned with the field. By applying pulses of RF power at the frequency particular to the nucleus under study (e.g. 64 MHz for protons in a 15 000 gauss magnetic field) the magnetization can be tilted away from the field. After termination of the pulse the magnetization tends to realign with the field with a time constant called the T_1 relaxation time. This process is analogous to the macroscopic situation of manually tilting a compass needle and watching it realign in a northerly direction. Throughout the 1950s and 1960s magnetic resonance was employed primarily in organic chemistry as T_1 and a second relaxation time called T_2 as well as the phenomenon of chemical shift were used to characterize the configurations of various molecules [3].

In the early 1970s the initial research was done in extending magnetic resonance to imaging [4]. The key step was to additionally apply a gradient of magnetic field to the sample, so that at different positions the resonant frequency is slightly different. In this way spatial encoding is performed, a necessary step in any kind of image formation. To use another analogy, this process is similar to having a pianist play a chord on a piano: by frequency analysis of the resultant sound wave it is possible to determine at what position the pianist struck the piano keys. By applying these concepts along the x-, y- and z-direction it is possible to determine the signal at each point, or, more conventionally, the signal as weighted by the extent to which the magnetization has recovered as dictated by the T_1 and T_2 phenomena.

The signal in a magnetic resonance (MR) image can be regarded as being dependent on both tissue-specific (intrinsic) and operator-selectable (extrinsic) parameters [5]. Intrinsic parameters include the spin density $N(H)$ and the spin-lattice and spin-spin relaxation times T_1 and T_2, which cannot be directly controlled by the experimenter but rather are characteristic of the materials being studied. Extrinsic parameters, which are operator-controlled, include the magnetic field strength, the specific sequence of pulses (inversion recovery, spin echo, etc.) and the times between the various pulses, commonly called the pulse delay and repetition times. Given a specific imaging system and pulse sequence, it is the delay and repetition times, in conjunction with the intrinsic parameters, that dictate the appearance of the final image. If the pulse sequence employed and the relaxation times of the two materials being compared are known, it is possible to calculate the delay and/or repetition times that will produce the maximum difference in signal intensity between those materials [5,6]. However, such techniques are somewhat limited, since the pulse delay time which is optimal for distinguishing materials A and B may not be optimal for differentiating A from C; thus the experimenter must decide which materials should be considered when selecting the pulse parameters. Moreover, the time required to obtain a single image is of the order of several minutes. For this reason it is not feasible from patient throughput considerations to indiscriminately take many scans while iteratively adjusting the scan parameters.

One approach to overcoming this problem is automated MR image synthesis, which allows rapid retrospective optimization of contrast via interactive control of scanning factors [7]. As shown in Fig. 1, MR image synthesis is a three-step process. First, multiple "source" images are acquired in the usual manner with the MR scanner. Next, "computed" images of the tissue properties (the proton density $N[H]$, T_1 and T_2) are formed by subjecting the source images to computerized "fitting" routines. Finally, "synthesized" images are generated and displayed by substituting operator-selected scanning factors (T_E, T_I and T_R) and the computed images into the equations that describe MR signal behaviour. This final step can be automated by using a high-speed digital video processor. Thus, with only one set of source images acquired with several combinations of pulse-sequence attributes, images can be synthesized for any desired scanning factors, enabling the viewer to optimize image contrast between selected tissues retrospectively.

In this paper we review the principles and applications of this technique.

Magnetic resonance image synthesis

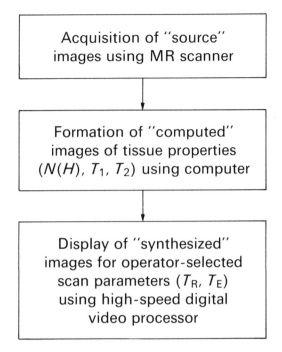

Fig. 1. Three steps of the NMR image-synthesis process.

2 DETERMINATION OF COMPUTED IMAGES

The first step in the MR image-synthesis process is the acquisition of source images. This is ideally done so that the T_1 and T_2 relaxation phenomena are sampled at several points. The pulse sequence that is usually used for this is called multiple spin-echo (MSE). At a given repetition time T_R the MR signal is measured at several different echo times T_E. These provide a measure for the T_2 relaxation. By repeating the process at a different T_R value the T_1 relaxation can also be estimated.

Typical values for the echo time T_E are in the range of 20 to 120 ms with four echoes at 25, 50, 75 and 100 ms being a common choice. Repetition times T_R are generally in the range 400 to 2000 ms. Total acquisition time is typically 256 T_R, or in excess of 8 minutes for a T_R of 2000 ms. The 256 factor occurs because each repetition provides the sampling along only one strip or line of Fourier space, so in general for $N \times N$ pixel resolution, N repetitions (or some multiple thereof) are required.

Figure 2 shows the measured signal S at an arbitrary pixel in the image plotted as a function of the T_E and T_R times used in the acquisition. The

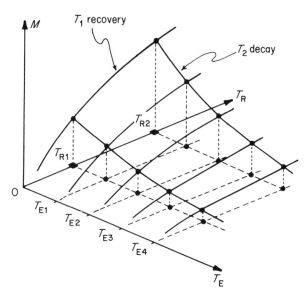

Fig. 2. Schematic diagram of the detected magnetization for an arbitrary pixel plotted as a function of the T_R and T_E values used in a multiple spin-echo (MSE) acquisition.

(T_E, T_R) points corresponding to the closed circles indicate the original MSE measurements. Note that with the MSE sequence described above four measurements are acquired for each T_R. For any fixed value of T_E the signal tends to increase as T_R increases, the rate of this increase being dictated by the longitudinal relaxation time T_1. Likewise, for any fixed value of T_R the measured signal tends to decay as T_E is increased, the decay constant being determined by the transverse relaxation time T_2. Thus the T_1 and T_2 relaxation phenomena can be considered independent and each can be determined separately. The final intrinsic parameter $N(H)$, or proton density, is proportional to the asymptotic value of the surface in Fig. 2 at $T_E = 0$ and large T_R.

To first order, at a given echo time such as T_{E1} the dependence of the signal S is given by

$$S(T_R, T_1) \propto 1 - e^{-T_R/T_1}. \tag{1}$$

By taking measurements at two values of T_R, e.g. T_{R1} and T_{R2} of Fig. 2, and forming the ratio R of these measurements, T_1 can be estimated by inverting the equation [8]:

$$R = \frac{S(T_{R1})}{S(T_{R2})} = \frac{1 - e^{-T_{R1}/T_1}}{1 - e^{-T_{R2}/T_1}}.$$

R decreases monotonically with increasing T_1. Improved accuracy and precision can be obtained either by incorporating measurements at additional T_R values [9] at the expense of scanning time, or by using the measurements at the additional echo times $T_{E2}, T_{E3}, \ldots$ [10].

The dependence of S on the other relaxation process T_2 can be modelled at fixed repetition time T_R according to

$$S(T_E, T_2) \approx PD\, e^{-T_E/T_2} \qquad (3)$$

where PD is defined as the pseudodensity and is equal to the value of S in the limit of $T_E = 0$. From a set of measurements $\{S_i\}$ acquired at echo times $\{T_{Ei}\}$ PD and T_2 can be estimated by first taking logarithms and then minimizing the quantity

$$R^2 = \sum_{i=1}^{N} [\ln S_i - (\ln PD - (T_{Ei}/T_2))]^2 \,. \qquad (4)$$

Variations are possible that incorporate weighted least-square minimization [11]. The results are estimates for $\ln PD$ and $1/T_2$ which best match the original data set.

Although it is outside the scope of this work, it is possible to optimize the values for the initial operator-selectable parameters T_R and T_E [11,12]. That is, from knowledge of the approximate range of values of T_1 and T_2 one can choose T_E and T_R that maximize the precision in the T_1 and T_2 estimates.

Finally, once T_1 and T_2 have been estimated, it is possible to use these estimates as well as the original measurements to estimate $N(H)$.

3 INSTRUMENTATION FOR RAPID MR IMAGE SYNTHESIS

Having discussed in the previous section the acquisition of source images and formation of computed images, we next discuss the final step of Fig. 1, the actual synthesis of images corresponding to arbitrary pulse parameters. This topic has been considered in detail by Lee et al. [13].

It is useful at the outset to state the requirements of the performance of the image synthesis device, before discussing the necessary hardware. First, the purpose of the device is to implement the third step described in Section 1, the synthesis step. This assumes that the basis quantities T_1, T_2 and $N(H)$ have already been computed for each pixel.

Secondly, it is assumed that for some given slice of interest the operator is to interactively combine the three basis images for that slice by manipulating the repetition and delay times T_R and T_E (or T_1). After trying several combinations of T_R and T_E, he decides which ones to use for the diagnostic images. From this it is clear that, once computed, the T_1, T_2 and $N(H)$ images for a slice may be used repeatedly by the operator. Hence these basis images should be stored in frame memories before the synthesis process begins, and recalled each time a timing parameter is modified. Also, it is evident that an interactive control, such as a joystick or trackball, is necessary to enable easy manipulation of the timing parameters.

Thirdly, the mathematical functions by which the basis images and operator-controlled timing parameters are to be combined are the equations describing the most commonly used pulse sequences: spin-echo (SE), inversion-recovery (IR) and saturation-recovery (SR). The signals S calculated for each of these sequences can be derived from the governing physical laws and are given explicitly by

SE: $\quad S = N(H)\,(1 + e^{-T_R/T_1} - 2e^{-(T_R - T_E/2)/T_1})\,e^{-T_E/T_2}$, (5 a)

IR: $\quad S = N(H)\,(1 + e^{-T_R/T_1} - 2e^{-T_1/T_1})$, (5 b)

SR: $\quad S = N(H)\,(1 - e^{-T_R/T_1})$. (5 c)

In order to generate the signals of (5a–c) rapidly enough for the synthesis process to be interactive, high-speed digital circuitry is required, ultimately resulting in a video image. A schematic of a device with the necessary hardware as dictated by these requirements is shown in Fig. 3.

The computations that must be performed in the digital circuitry are evident in the above pulse-sequence equations (5 a–c). Addition, subtraction and multiplication can be carried out in arithmetic logic units (ALUs) and digital multipliers. The exponential terms require reciprocation of T_1 or T_2, multiplication by the appropriate interpulse time, and exponentiation.

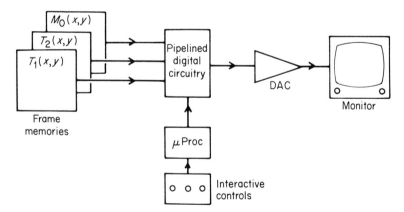

Fig. 3. Schematic diagram of the essential components of an image processor capable of rapid MR image synthesis.

Reciprocation can either be done during synthesis, or performed once before synthesis is begun, and the resultant $1/T_1$ and $1/T_2$ images stored in frame memories rather than T_1 and T_2. Exponentiation is readily accomplished with look-up tables (LUT). Each of these components—ALUs, multipliers and LUTs—can operate at the rate required (~ 10 MHz) to process a 512×512 image within a $1/30$ s video frame interval. Addition, subtraction and multiplication can be performed without introducing error when bit-limited hardware is used. Reciprocals and exponentials, however, cannot be calculated perfectly with bit-limited hardware, and will therefore introduce some error into the synthesis calculations. Although it is outside the scope of this work, it is possible to show [13] that the use of 16-bit processing introduces, owing to computational precision, an error that is at most 0.4% of that in the maximum MR signal $N(H)$. This is generally well within that due to the statistical uncertainties or noise in the original measurements.

It is possible to design a network of digital components in which the signal S of any of (5 a–c) can be generated for all pixels in a 512×512 image within a $1/30$ s video frame interval. As an alternative, and with a slight increase in response time, we have developed a method for performing rapid MR image synthesis using a commercially available digital image processor (IP8400, Gould, Inc., DeAnza Imaging, San Jose, CA) driven by a minicomputer (LSI 11/23, Digital Equipment Corp., Maynard, MA). The processor consists of twelve 512×512 pixel 8-bit frame memories, a joystick for entering new values of T_E (or T_I) and T_R, and a pipelined digital video processor (DVP). A simplified schematic of the DVP is shown in Fig. 4. As shown, the contents of up to four frame memories can be simultaneously passed through and manipulated within the DVP in one video frame interval.

Because the DVP of Fig. 4 cannot generate the complete expression of (5 a–c) in one pass, it is used to generate one term at a time. As an example, suppose that saturation-recovery images (5 c) were to be synthetically

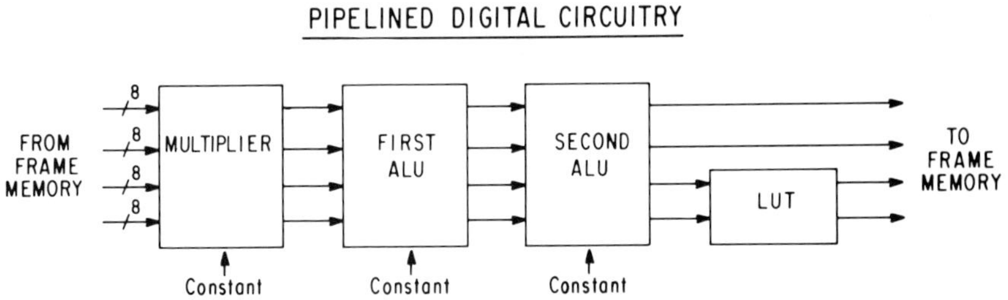

Fig. 4. Simplified schematic of Gould DeAnza IP8400 Digital Video Processor.

Table 1
Sequence of steps used to synthesize saturation-recovery images

Video frame	DVP function	Memory contents $M_0:M_1$	$M_2:M_3$	$M_4:M_5$	M_6
		$N(H)$	$1/T_1$		
1	Multiply	$N(H)$	$1/T_1$	T_R/T_1	
2	LUT	$N(H)$	$1/T_1$	$(1-e^{-T_R/T_1})$	
3	Multiply	$N(H)$	$1/T_1$	same	$N(H)(1-e^{-T_R/T_1})$

formed. The sequence of steps used to form such an image is shown in Table 1. Prior to enabling the operator to manipulate T_R, the proton density image $N(H)$ is stored at 16-bit precision in two frame memories, M_0 and M_1. The T_1 image is reciprocated and the resultant $1/T_1$ image stored in memories M_2 and M_3. These memory contents remain unaltered for all subsequent steps.

Next, the synthesis process begins. During the first video frame $1/T_1$ is read from memories $M_2:M_3$, and multiplied within the DVP by the 12-bit constant T_R. Sixteen bits of the possible 28-bit product are stored in additional memories $M_4:M_5$. During the second frame the T_R/T_1 image just formed is read from $M_4:M_5$ and the exponential $1-e^{-T_R/T_1}$ is formed at every pixel using the DVP LUT. During this frame the pixel values are passed through the multiplier and ALU stages without alteration. The result is again stored in memories $M_4:M_5$. In the third pass this result is multiplied by $N(H)$, yielding the desired signal S for every pixel. During this third frame the final result is stored in an additional memory M_6. For all video frame intervals the contents of M_6 are converted into analogue video

Table 2
Characteristics of processing for MR image synthesis

Pulse sequence	Computed images	Operator-controlled parameters	DVP passes
Simple spin-echo	Pseudodensity, T_2	T_E	3
Full spin-echo	$N(H)$, T_1, T_2	T_E, T_R	9
Saturation recovery	$N(H)$, T_1	T_R	3
Inversion recovery (real or absolute value)	$N(H)$, T_1	T_I, T_R	8

format to refresh a monitor. Upon seeing this new SR image the operator can alter T_R via the joystick and the process will repeat.

It should be clear that the method discussed in this SR example can be extended to the spin-echo and inversion-recovery pulse sequences. A more accurate model using multi-exponential relaxation could also be incorporated. In either case, the more complex pulse-sequence equations could require more memories in which to store intermediate results, and more passes through the DVP to complete the calculation. Characteristics of synthetic processing for various pulse sequences are summarized in Table 2.

4 VALIDATION STUDIES

Having discussed the formation of computed images of T_1, T_2 and $N(H)$, and the instrumentation that can be used for image synthesis, perhaps the first question one asks is whether synthetic images truly represent those which are directly acquired. We have attempted to study this question in a series of comparisons having an increasing degree of rigour. In describing these experiments we refer to Fig. 2, which shows the T_E and T_R values at which spin-echo images are acquired and synthesized.

4.1 Synthesis of images originally acquired

The first test was to take images at several different echo times, apply a weighted least-squares fit to them to determine T_2 and $N(H)$, and then use these fitted parameters to synthesize images at the original echo times. We have shown that it is possible to consistently reproduce the original images to an accuracy within the noise level of the original measurements themselves [14]. This is illustrated in Fig. 5, a plot of the acquired and synthetic signals in a 25 pixel region of interest (ROI) shown versus the echo time T_E. The same T_E was used for each pair of signals plotted. The vertical spread in each case corresponds simply to the statistics in sampling the same material 25 times. As shown, the average synthetic signal (the midpoint of the spread of synthetic values for a particular T_E) is consistently within the noise or spread of the original signals.

4.2 Interpolation/extrapolation in T_E

The next test was to take data at several echo times (24, 48 and 68 ms), perform the T_2 and $N(H)$ fits, and resynthesize at a different echo time (92

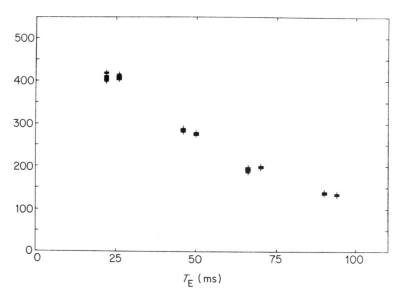

Fig. 5. Plots of NMR signal vs. echo time for 5 × 5 pixel regions of interest in cerebrum. For each pair of signals the source values are on the left and the synthetic values are on the right. Each horizontal bar represents the signal for a single pixel.

ms). This result was compared with an image directly acquired at this echo time. Thus the synthetic image is being compared with an actual image that was not used in the T_2, $N(H)$ fit. Such a comparison is shown in Fig. 6, and again is very favourable. Similar results are obtained if instead of extrapolation one performs an interpolation in T_E.

4.3 Interpolation/extrapolation in T_R

The tests discussed thus far and similar ones have shown that it is valid to synthesize images for T_E times different from those of the acquisition. The third validation test was to see if interpolation or extrapolation along the T_R direction was valid. This was done by first acquiring data at T_{R1} and T_{R2} of Fig. 2 (e.g. 500 and 2000 ms) for the same echo time T_E. Next, as described in Section 2, T_1 and a density were estimated. Finally these were inserted back into the appropriate equation and an image was synthesized for an alternative T_R value. This result was then compared with an image acquired directly at that T_R. As in the earlier tests, the correspondence between synthetic and acquired images was very close. In fact, in some cases the synthetic result was appreciably less noisy than its acquired counterpart.

Magnetic resonance image synthesis

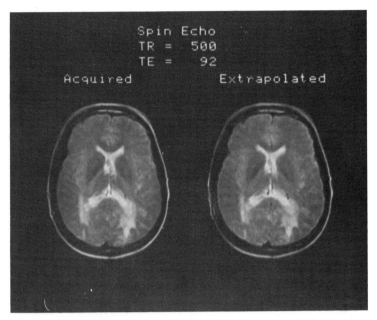

Fig. 6. Acquired and synthetic spin-echo images of the same slice of a patient having necrosis secondary to radiation therapy for an astrocytoma. The synthetic result was extrapolated to a T_E of 92 ms from acquired images at 24, 48 and 68 ms.

4.4 Simultaneous interpolation/extrapolation in T_E/T_R

To permit the generation of arbitrary synthetic spin-echo images it is necessary to allow interpolation and extrapolation in both T_E and T_R. This was studied as follows [15]. First one determines the PD and T_2 images for each of the T_R times used. Next, as described in the previous test, one moves along the T_R axis by incorporating the PD images into a T_1 fit. Finally, after going to the desired T_R time, one moves out along the T_E direction by using the T_2 image calculated in the first step. Thus arbitrary values of T_R and T_E are allowed. A comparison of a synthetic image interpolated in both T_E and T_R is shown in Fig. 7 and compared with an image directly acquired using the same pulse parameters. Again, the comparison is excellent, both in the accuracy with which the synthetic signals match those of the direct acquisition and in the comparable noise levels.

We conclude from these experiments that synthetic images do in fact closely match those directly acquired in both accuracy and precision.

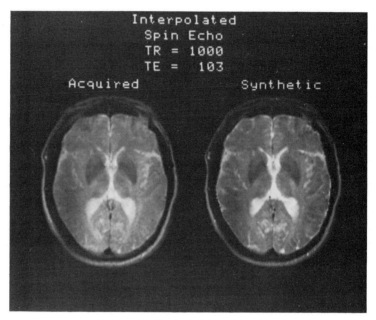

Fig. 7. Acquired and synthetically interpolated spin-echo images of a transaxial section of the head of a patient with suspected hemi-Parkinsonism for a repetition time T_R of 1000 ms and echo time T_E of 103 ms. The synthetic image was generated from source data acquired at T_R values of 500 and 2000 ms.

5 APPLICATIONS

The motivation for performing MR image synthesis was to enable the simulation of any set of pulse-sequence parameters after completion of the scan. The previous sections have discussed the concepts and instrumentation which enable this. In this section we present three additional specific applications.

5.1 Pulse-sequence extrapolation

Thus far the discussion has focused on the multiple spin-echo pulse sequence, one that is efficient in the sense that several images are acquired in the time normally required for only one. This sequence, however, provides intrinsically less contrast between materials that have slightly different T_1 times than do other sequences, most notably inversion recovery (IR) [16]. The disadvantage of IR acquisition, however, is that it is very time-consuming.

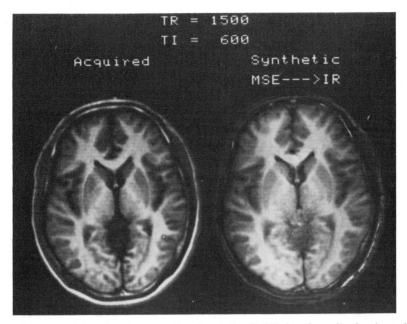

Fig. 8. Transaxial section of a normal head at the level of the basal ganglia showing pairs of inversion-recovery images with $T_R = 1500$ ms, $T_E = 27$ ms and $T_I = 600$ ms. The image on the left was experimentally acquired; the one on the right was synthetically extrapolated from MSE data.

The purpose of pulse-sequence extrapolation is essentially to have the best of both worlds: temporally efficient acquisition and high T_1-based contrast. This technique has been studied by Bobman et al. [17]. It is accomplished by acquiring data with MSE acquisitions at two T_R values, as depicted in Fig. 2, performing the regressions to obtain T_1, T_2 and $N(H)$, and then inserting them into the IR signal equation (5b). During the synthesis step the inversion time T_I can be interactively adjusted to yield images whose T_1-based signal can vary markedly. A comparison of a synthetic IR image with one directly acquired is shown in Fig. 8.

5.2 SNR enhancement

Every synthetic image can be considered in a qualitative sense as an "average" of the multiple acquired images used to form it. Although the average is not a strict arithmetic one, nevertheless one may expect to obtain with synthesis some of the advantages attendant to averaging, the most common being noise reduction. This in fact happens when, at fixed T_R, spin-echo images are synthesized at alternate values of the echo time T_E.

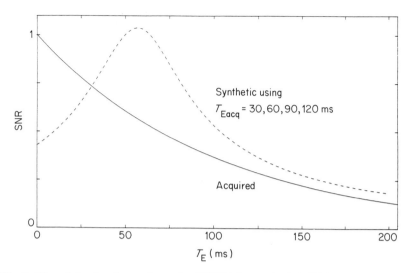

Fig. 9. Plot of the signal-to-noise ratios (SNR) in acquired (solid line) and synthetic (dashed line) MR signals assuming echoes were originally acquired at 30, 60, 90 and 120 ms. For synthetic echo times beyond approximately 30 ms the synthetic result is superior.

Referring again to Fig. 2, suppose images are acquired at T_{R1} at echo times $T_{E1}, \ldots, T_{E4}$. Suppose next that these signals are used to synthesize a new image at T_{Es}, which is greater than T_{E1}. Because the four echoes used to synthesize the synthetic signal have signal levels themselves that are comparable to or exceed the new signal, and because some noise averaging is expected, one expects the signal-to-noise ratio (SNR) in the synthetic result to exceed that in the acquired signal. In fact this has been treated rigorously by MacFall et al. [11], and sample results are shown in Fig. 9. The plot shows that for T_E values exceeding approximately the first echo time the synthetic SNR is greater than that in acquired signals. In the vicinity of the second echo it can be almost two times greater. Even for late echo times which are often important for visualizing pathology, the synthetic result is superior.

5.3 Scan-time reduction

One may question whether scanning time can be reduced via the image-synthesis technique. This would be done by acquiring data at repetition times T_{R1} and T_{R2} and synthesizing at a third time $T_R > T_{R1} + T_{R2}$. One expects intuitively that beyond a certain point such a process will

Magnetic resonance image synthesis

substantially amplify statistical noise, substantially limiting the utility of the method.

This process has been studied in detail by Lee et al. [12], and typical results are shown in Fig. 10. Plotted there is a quantity called NA, the noise amplification factor, versus the total net T_R required for the synthesis process. This assumes that an image is subsequently synthesized from this data at a T_R of 2000 ms. NA is defined as the ratio of standard deviations in

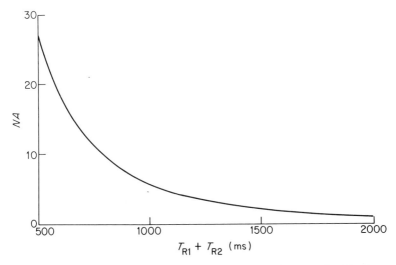

Fig. 10. Plot of the noise amplification NA in a signal synthesized at a T_R of 2000 ms and $T_1 = 800$ ms from data acquired at the total repetition shown. For a total acquisition T_R of 1600 ms (20% time reduction versus 2000 ms) the noise level increases approximately twofold.

the noise level in the synthetic image vs. that in an image directly acquired at 2000 ms. As shown, for a 20% reduction in scan time the noise level increases by about a factor of 2. Such a factor can potentially be compensated in part by noise averaging in the orthogonal T_E direction. For time reduction in excess of 20%, however, the noise amplification becomes quite severe.

Although scan-time reduction does not thus appear to be a major application of image synthesis, it should be pointed out that in a related application Lee [12] has shown that one can synthesize for any T_R up to $T_{R1} + T_{R2}$ with at most a 20% noise increase. Thus any "penalty" for having the flexibility of image synthesis is very modest.

6 SUMMARY

We have discussed some of the limitations of MR imaging to high patient throughput. As a solution to these limits we have introduced the concept of MR image synthesis. We have discussed how the computed images of T_1, T_2 and $N(H)$ can be formed in an optimum fashion and how the synthesis step can be automated with high-speed digital instrumentation. Synthetic images closely match acquired ones in accuracy and precision. The method can thus be used to address the limitation that motivated its development as well as in a number of other applications.

ACKNOWLEDGMENTS

This investigation was supported by PHS grant 1 R01 CA37993, awarded by the National Cancer Institute, DHHS; General Electric Medical Systems; the Whitaker Foundation; and Gould/DeAnza Imaging.

REFERENCES

[1] Purcell, E. M., Torrey, H. C. and Pound, R. V. (1946). Resonance absorption by nuclear magnetic moments in a solid. *Phys. Rev.* **69**, 37–38.
[2] Bloch, F. (1946). Nuclear induction. *Phys. Rev.* **70**, 460–473.
[3] Farrar, T. C. and Becker, E. D. (1971). *Pulse and Fourier Transform NMR*. Academic Press, New York.
[4] Lauterbur, P. C. (1973). Image formation by induced local interactions: examples employing nuclear magnetic resonance. *Nature* **242**, 190–191.
[5] Wehrli, F. W., MacFall, J. R. and Glover, G. H. (1983). The dependence of nuclear magnetic resonance (NMR) image contrast on intrinsic and operator-selectable parameters. *Proc. SPIE* **419**, 256–264.
[6] Wehrli, F. W., MacFall, J. R. and Newton, T. H. (1983). Parameters determining the appearance of NMR images. In *Modern Neuroradiology Advanced Imaging Techniques*, Vol. 2 (ed. T. H. Newton and D. G. Potts), pp. 81–117. Claradel Press, San Anselmo.
[7] Riederer, S. J., Suddarth, S. A., Bobman, S. A., Lee, J. N., Wang, H. Z. and MacFall, J. R. (1984). Automated MR image synthesis: feasibility studies. *Radiology* **153**, 203–206.
[8] Lin, M. S. (1984). Measurement of spin-lattice relaxation times in double spin-echo imaging. *Mag. Res. in Med.* **1**, 361–369.
[9] Liepert, D. W. and Marquardt, D. W. (1976). Statistical analysis of NMR spin-lattice relaxation times. *J. Mag. Res.* **24**, 181–199.
[10] Riederer, S. J., Bobman, S. A., Lee, J. N., Farzaneh, F. and Wang, H. Z. Improved precision in calculated T1 images using multiple spin-echo acquisition. *J. Comp. Assist. Tomogr.* (submitted).
[11] MacFall, J. R., Riederer, S. J. and Wang H. Z. An analysis of noise propagation

in computed T2 and pseudo-density and synthetic spin-echo images. *Med. Phys.* in press.
[12] Lee, J. N., Bobman, S. A., Johnson, J. P., Farzaneh, F. and Riederer, S. J. (1986). Noise propagation in TR-extrapolated synthetic images. *Med. Phys.* **13,** 170–176.
[13] Lee, J. N., Riederer, S. J., Bobman, S. A., Farzaneh, F. and Wang, H. Z. (1986). Instrumentation for rapid MR image synthesis. *Mag. Res. in Med.* **3,** 33–43.
[14] Bobman, S. A., Riederer, S. J., Lee, J. N., Suddarth, S. A., Wang, H. Z., Drayer, B. P. and MacFall, J. R. (1985). Cerebral MR image synthesis. *Am. J. Neurorad.* **6,** 265–269.
[15] Bobman, S. A., Riederer, S. J., Lee, J. N., Suddarth, S. A., Wang, H. Z. and MacFall, J. R. (1985). Synthesized MR images: comparison with acquired images. *Radiology* **155,** 731–738.
[16] Edelstein, W. A., Bottomley, P. A., Hart, H. R. and Smith, L. S. (1983). Signal, noise, and contrast in nuclear magnetic resonance (NMR) imaging. *J. Comp. Assist. Tomogr.* **7,** 391–401.
[17] Bobman, S. A., Riederer, S. J., Lee, J. N., Farzaneh, F. and Wang, H. Z. (1986). Pulse-sequence extrapolation using MR image synthesis. *Radiology* **159,** 253–258.

CONCLUSION

A multi-author book has a tendency to reach the end of its last chapter and then just stop. It would indeed be pleasant to be able to write a scholarly resumé of the eighteen chapters forming this volume, but, not surprisingly perhaps, this would appear to be an almost impossible task. Each contribution is in itself a summary of its author's thoughts in the loosely defined area of intermediate-level image processing; any further summarizing would be liable to cause the essential points being made to be lost.

It is the editor's privilege to include material in a collection such as this, which would otherwise have little chance of being published. At the risk of being accused of abuse of privilege, the editor has therefore added a tailpiece in lieu of a summary in the hope that a little unfounded speculation will not be seen as a drowning researcher clutching at a straw.

Chapter Nineteen
Complexity

M. J. B. Duff

1 INTRODUCTION

The methods of scientific discovery appear to work satisfactorily most of the time. Implicit within them there is a fourteenth-century principle, which even today would seem to be inviolable:

the simplest explanation is probably the right one and should therefore always be adopted

(William of Ockham, 1280–1348, stated this rather more elegantly although, perhaps, more obscurely, in the form:

entia non sunt multiplicanda praeter necessitatem

—entities ought not to be multiplied except from necessity.)

In this closing chapter of a volume devoted to the application of current scientific methods to the task of analysing image data, it may be profitable to speculate gently on the hypothesis that simplicity need not always be the best guide, particularly in the quest for faster and more efficient computing structures.

Early attempts to process images with computers were constrained by resources available at the time, these implying the use of conventional von Neumann machines programmed, more often than not, in FORTRAN. During the same period, physiologists, anatomists and psychologists were aware of and often deeply interested in the immensely complex and highly effective biological image-analysis systems known to lie behind the mammalian eye. Linking these two otherwise disjoint fields of study was a relatively small group of "fringe scientists" who were trying to design neuron models in hardware or software, hoping eventually to simulate neural nets in order both to understand them and to predict their behaviour.

It is not unreasonable to ask whether these activities led directly to today's special-purpose image-analysis computers. Do SIMD processor arrays, pipelines, systolic arrays, bus-structured multiprocessors, and so on, owe anything to their biological predecessors? Certainly, the ubiquitous local neighbourhood processor architectures call to mind the on- or off-centre retinal fields defined by neurons in the visual cortex and considered to act as orientated line-segment detectors. Clearly, also, the processor-per-pixel SIMD arrays bear some resemblance to the structure of the retina. However, beyond these somewhat superficial comparisons, points of similarity are hard to find. On the whole, it would seem that computer architectures have not been greatly influenced by studies of living organisms.

In contrast, the art of algorithm design has been almost unhealthily introspective. The natural inclination for a new graduate student asked to solve an image-analysis problem is always to rush headlong into a solution without so much as a glance at the literature. Each of us is aware of being able both to see and interpret images with virtually no conscious effort; all that has to be done, therefore, is to think out how we do it, or so it would seem. In practice, the intuitive approach is fickle: it sometimes works but it often fails us. When it does, there would appear to be no alternative but merely to think again.

2 THE PROBLEM

An ideal image-analysis system would be cheap, small, economical in its power consumption, easy to program, self-contained (and/or easily interfaceable) and, above all, capable of performing rapid image analysis. The extent to which inevitable compromise will be allowed to sacrifice one or more of these ideals in favour of another is a matter of expediency. It is always possible to improve performance, if only marginally, by spending more money, although the resulting monstrosity well might overflow the laboratory. But this is not the point of the present discussion; we are concerned to know why existing efficient machines are particularly efficient and to decide in what direction we should direct our attention in order to progress to increased efficiency. Merely increasing the clock rate of the semiconductors or doubling up the number of processors (the "expensive" approach) is neither interesting nor very rewarding. A closer examination of images and architectures and the relationships between them should be more profitable.

Whether we like it or not, image data is profuse. A picture may be "worth a thousand words" but a 512 by 512, 8-bit pixel image fills 131 072 words in the memory of a 16-bit word-length computer. Image analysis in its early stages largely consists in attempting to throw away a substantial fraction of this mainly redundant data, retaining only those parts that contain important information. The key to the almost insuperable difficulty in doing so is hidden in the word "important". What is "important" in an image will depend on the purpose for which the analysis is being carried out–which is why it is foolish to expect to be able to build an image analyser of general applicability and high performance over a wide range of tasks. It is also the reason why so-called knowledge-based methods are increasingly being proposed.

With hindsight, it is easy to see how special-purpose image computers have been designed in order to achieve efficiency. Two fundamental principles are involved: first, the processor and/or memory structures should match the two-dimensional structure of image data, and, secondly, they also should match or relate organizationally to the more commonly occurring image-processing functions. Examples demonstrating these two propositions can be found in abundance in this and earlier related volumes. Unfortunately, the data and function structures characteristic of low-level image processing are replaced by very different structures at higher levels. It is then no longer necessarily true that the reduced data will be organized as a two-dimensional array, and, similarly, the processing operations appropriate to high-level data formats (feature strength lists, for example) will also be of a very different shape. In particular, the local neighbourhood operator will almost certainly have lost its relevance.

If we nevertheless still try to apply the principle of matching discussed above, a further complication is encountered: not only has the two-dimensional data array disappeared but the new data structures take a variety of forms, depending on the problem. Thus one process might generate a grey-level histogram while another will result in a chain code. Some processes construct higher-dimensional data sets (such as three-dimensional models), whereas others reduce data to a few parameters or even to a yes/no decision. It would obviously be impossible simultaneously to match processor architectures to these widely varying data sets.

3 A NEW HYPOTHESIS

Faced with these difficulties, it is prudent to re-examine our adopted guidelines. Is it possible that processing power might be improved by approaching the problem from a completely different standpoint?

In fact, there is another property of existing image-processing architectures that happens to have accompanied the matching principle: complexity. In some systems the complexity appears in the simultaneous application to the data of large arrays of simple processors; in others, it is the data access mechanism that is multiplied up (a "parallel fetch"). Perhaps, therefore, a new hypothesis should be subjected to careful examination:

> unpredictable properties emerge from systems whose complexity is higher than necessity demands

In human terms, this hypothesis is tacitly accepted. The saying that "two heads are better than one" is not meant to imply only that problems are solved twice as quickly by two people working together, but rather that some intractable problems might never be conquered by a lone thinker. On a more mundane level, many tasks about the house seem to need at least three hands, and, in research, it is often the case that a large research group will offer opportunities for cross-fertilization of ideas leading to a research output far exceeding in quantity and quality that to be expected from the individuals involved.

The analogy can be taken further. Too large a group can become uncontrollable and self-defeating; all its energies are taken up in maintaining the integrity of the whole. This process was described, tongue-in-cheek, by C. Northcote Parkinson in one of the extensions to his notorious "Law". The computer analogue of this phenomenon is the ill-designed multiprocessor in which most of the operations are data and instruction transfers between the subsystems. Adding more processors serves only to aggravate the internal traffic load.

4 APPLYING THE HYPOTHESIS

Starting, therefore, with the intention of increasing system complexity (ultimately with a view to improving performance), there are several ways to proceed. Whichever path is taken, sight must not be lost of the fact that the complex system has to be controllable or programmable so that it can be made to perform the required operations. This requirement might seem to be self-evident, but circuits effecting unexpected and interesting picture transformations are sometimes wrongly categorized as image computers. The chance of this occurring becomes greater as the system complexity rises.

Complexity can be measured, or at least described, in terms of the following three parameters:

(i) a program complexity number P, which is a measure of the number

Complexity

of independent PE (processing element) control programs simultaneously executable;
(ii) an internal complexity number I, which is a measure of the number of usefully active circuit elements within each individual PE;
(iii) a connectivity complexity number C, which is a measure of the number of connections made by each PE (to memory or to other PEs).

An overall complexity can be defined which is merely the product PIC, but to suggest that this would be a numerically significant figure of merit would be absurd and certainly not supportable by experimental evidence. The thrice repeated phrase: "a measure of" underlines the danger of resorting too quickly to numbers in this sort of exercise (a tendency also disturbingly to be found amongst benchmarkers). However, cautiously adopting the PIC philosophy whilst avoiding its numerical implications, it would seem indisputable that a maximally complex system would be one in which there was a vast network of sophisticated PEs, each connected to all the others and each capable of executing its own program for a substantial fraction of the time. The trouble is that such a system would probably be impossible either to build or to program using present day skills.

So, to a first approximation, let us identify two broad areas of research to be pursued if we wish to achieve high PIC ratings: connections and multiple programming.

5 FURTHER RESEARCH

5.1 Connections

In a two-dimensional image-data-related architecture, it is now conventional to make connections between processors and their immediate 3 by 3 local neighbours. This connectivity supports all image processes (as sequences of 3 by 3 neighbourhood operations), but the support becomes seriously weaker when larger neighbourhood operators are required or when data need to be moved over long distances in the array. Multiresolution analysis is particularly badly served, and pyramid architectures have been proposed to rectify this omission. Another difficulty is that found when transferring data to and from the array. At higher levels, semantic nets would appear to demand connection schemes of much higher dimensionality than 2.

Solutions that have been proposed in the past mostly suffer from the same deficiency: they will not support the simultaneous transmission of data between unrelated pairs of sources and receivers (much in the same way as two RAM addresses cannot be accessed at the same time or as a bus cannot carry two messages concurrently). It would seem that the only way out of the difficulty is to hard-wire connections between all pairs of communicating terminals. But is there another factor to be taken into account that might lessen the magnitude of this task? In fact, it would seem so. Low-level data (e.g. pixels) are locally related; high-level data are non-locally related. Expressed another way, we can carry out operations on images at the pixel level, involving only pixels that are grouped together, whereas analysis of a feature set may well involve relating pairs of features drawn from any regions of the set. The implication of this line of reasoning is that the amount of interconnection between processors should increase as the data processed by them become more abstracted. Since, at the same time, the number of data elements decreases, the overall pattern is of an exchange between the numbers of processors in use and the number of connecting paths between them. Finally, there is some evidence to suggest that the operations required at the higher levels are more complex than those at the lower levels.

The picture that emerges is of a form of pyramid in which interprocessor connectivity and processor complexity both increase in passing from lower to higher levels. At every level, an additional structure could be superimposed to facilitate the rapid bulk transfer of data into or out of the level, either between adjacent layers or to and from outside the pyramid.

5.2 Multiple programming

An image-processing task is usually split up in one of three ways:

(a) the SIMD method, in which processors are assigned to particular regions of the image (the regions may even be as small as single pixels) and in which the complete process is implemented as a sequence of subprocesses performed simultaneously and identically by every processor;

(b) the pipeline method, in which streams of pixels or complete images are channelled through a chain of processors, each of which performs repeatedly the same operation on successive pixels or images;

(c) the MIMD method, in which an assembly of identical or special-purpose processors, often connected via a high-speed bus, operates on various sections of the data and of the task, under the general supervision of a master controller.

Considering the low-level processing implied in (a), it can be appreciated that many of the tasks to be performed are easily expressed but hard to

Complexity 313

program in an SIMD mode. For example, estimating the areas of overlapping spherical blobs (circular in cross-section) is a difficult task apparently requiring the extrapolation of the blob perimeters. A related problem is that of interpolating between the segments of a broken, non-straight line. Yet another is the definition of region boundaries in an image containing a wide variety of textures. All these and many other tasks could be performed more efficiently if processors could be locally programmed in such a way that the programs were capable of being strongly influenced by local data. In order to achieve a substantial gain, it would be necessary to do more than divide the processors into a few subgroups, each group running under an SIMD program for all the processors within the group. To a first approximation, this would give a gain only of the order of the number of groups. Worthwhile gains should arise when every processor is, in principle, able to be diverted onto its own program; the gain might then be of the order of the number of processors—presumably much higher than the number of groups. Note that although superficially similar, the structure to support this principle is not a dataflow architecture. However, a step in this direction has been taken in the proposed design for CLIP7.

At higher levels, where the number of processors is very much smaller and the processor complexity potentially higher, it becomes reasonable to propose that each processor should have its own program store and controller. The possibility of passing program segments between processors should also not be ignored. This strategy corresponds to a human situation in which a task, or part of it, is delegated to an assistant.

6 CONCLUSIONS

Ockham's Razor may not be as sharp as it used to be. A new hypothesis suggests that powerful image-processing architectures might evolve if a deliberate attempt were to be made to increase system complexity for its own sake. It is further suggested that the principle might be explored in a layered, pyramid-like structure in which local neighbourhood connections in the lowest layer are increased to multidimensional interconnection in the highest levels, at the same time increasing processor complexity. Again, while in the lowest layer programming would be basically SIMD, the program could be locally modified by local data. The same principle would apply at higher levels, eventually permitting the transfer of program segments between the processors. Whether or not a system of this type would adequately support the type of algorithms image analysts want to employ has not been established. It could be that the high connectivity and high PE power in the upper layers of the pyramid would not get over the weakness of the less dense connectivity lower down. Only a careful study of representative algorithms and trial structures would throw light on this problem; this discussion does little more than suggest an approach which might be worth pursuing. The system tentatively proposed certainly achieves one objective: it is excessively complex. It would be interesting to see if such a design would also support the complexity hypothesis by showing itself to be an efficient image processor.

Index

Address calculation, 90
Algorithms, near neighbour, 204–207
  parallel, 53 et seq.
  task flow of, 112–115
  time complexity of, 107–110
  virtual, 54–56
Applications, chromosome analysis, 11
  histopathologic sections, 268–272
  liver cell analysis, 248–251
  magnetic resonance image synthesis, 287 et seq.
  medical images, 267 et seq.
  road extraction, 87–88
  thyroid, 269–272
Architecture model, 56–57

Binocular vision, 137–138

Chain-run encoder, 12
Communication diameter, 154
Complexity, 307 et seq.
Computed images, 287 et seq.
Computers, AAP, 66
  augmented pyramid, 131 et seq.
  Butterfly, 211
  C.mmp, 211
  CHiP, 66
  CLIP, 3, 7, 132–133, 209–210, 234–235, 256, 259
  Cosmic Cube, 66
  DADO, 209, 211
  DAP, 132–133, 234
  Gould DeAnza, 293–295
  hypercubes, 161–163
  I-ocular pyramids, 139
  iconic, 3 et seq.
  ILLIAC III, 132
  ILLIAC IV, 210

inner product, 258
Intel 432, 198–208
ISMAP, 11–12
LISP machine, 4
Magiscan M2, 24–33
mesh, 149 et seq.,
MPP pyramid, 177–179
MPP, 3, 7, 66, 132–133, 209–210
Multicluster, 193 et seq.
multimodal pyramids, 140–143
NYU Ultracomputer, 103, 211
PAPIA, 174–177, 188–189
PASM, 7, 209 et seq.
PCLA, 172–175
PIPE, 12
POLYP, 272–285
pyramids, 131 et seq., 149 et seq., 167 et seq.
RP-3, 211
SNAP, 13
SOLOMON-I, 132
STARAN, 210
SYMPATI, 231 et seq.
systolic, 259
TRAC, 211
VAP, 243 et seq.
VISIONS, 5
VM1, 24
ZMOB, 211
Connected component labelling, 101 et seq.
Connectivity, 11
Contour labelling, 181 et seq.
Convolution, 26–27

Diffusion process, 182–189

Execution profile, 203–204

315

Feature extraction, 36 et seq.
Flexible-slit method, 36 et seq.
Focus-of-attention, 21–22, 103

Gallium arsenide, 253 et seq.

Heuristic image operations, 93–94
Hit function, 10
Hough transform, 9–11

Iconic-to-symbolic transformation, 4–16
Image sequences, 139–140
Image-description system, 47
Interconnection network, 215, 217, 219–221, 232
Interfaces, 13–14
Interprocess communication, 72, 280–283
Interprocessor communication, 154–163, 311–312

Languages, C, 26, 67 et seq.
  LOOPS, 33
  p-code, 25–27
  Parallel PASCAL, 66
  PASCAL, 25–26
  Poker, 66
  shape description, 36, 48
  V-language, 65 et seq.
Laser scanner microscope
Look-up tables, 246–248

Macropipelining, 93
Memory, address processor, 29–30
  bimodal, 14
Message passing, 226–227
Microcode, 90
Microprogram control, 27
Mode switching, 224–225
Morphosis, 65

Multifunction processing, 209 et seq.
Multilevel structures, 231 et seq.
Multiple programming, 312–313

Neighbourhood operations, 26–27

Ockham's Razor, 307

Parallel iconic operations, 95
Processing levels, 92–93
Projection curves, 37 et seq.
Propagation, 236–237
Pyramid performance, 159–160

Reconfigurable parallel architectures, 53 et seq.
Region growing, 86
Replicator, 168–169
Retinotopic architectures, 4–5
Rule-based system, 239–240

Segmentation of aerial images, 85
Selector, 168–169
Shared data structures, 21
Shared memory, 226–227
Slider, 168–169
Spin-echo images, 297–298
Structured process, 67–80
SUM-OR, 171–172
Symbolic computation, 4
Symbolic-to-iconic transformations, 5
Synchronization, 104, 115–116

Task distribution, 193 et seq.
Task flow analysis, 104
Touch, 141

VLSI, 253 et seq.